ON THE ANTIQUITY

OF INTELLECTUAL MAN

ON THE ANTIQUITY OF INTELLECTUAL MAN

FROM A PRACTICAL AND ASTRONOMICAL POINT OF VIEW

BY

C. PIAZZI SMYTH, F.R.SS. L & E,

F.R.A.S, F.R.S.S.A, HON M. INST. ENG. SC., PHOT. S. ED., AND R. A. A.S. MUNICH A PALERMO; PROFESSOR OF PRACTICAL ASTRONOMY IN THE UNIVERSITY OF EDINSURGH, AND ASTRONOMER-ROYAL FOR SCOTLAND.

ISBN: 978-93-5128-750-6(PB)

First Published in 1868
Indian Reprint in 2017

Published by

Kalpaz Publications
C-30, Satyawati Nagar,
Delhi – 110052
E-mail: kalpaz@hotmail.com
Website: kalpazpublications.com
Ph.: +91-9212142040

Cataloging in Publication Data—DK
Courtesy: D.K. Agencies (P) Ltd. <docinfo@dkagencies.com>

Smyth, C. Piazzi (Charles Piazzi), 1819-1900, **author.**
On the antiquity of intellectual man, from a practical and astronomical point of view / by C. Piazzi Smyth (F.R.SS. L. & E.).
pages cm
Reprint. Originally published: Edinburgh : Edmonston and Douglas, 1868.
"The present little work may be regarded as a sequel to the author's recent volumes on the Great Pyramid of Egypt"—Preface.
ISBN 9789351287506

1. Architecture, Ancient. 2. Pyramids. 3. Civilization, Ancient. I. Title.

DT63.S69 2017 DDC 932.012 23

PREFACE.

THE present little work may be regarded as a sequel to the author's recent volumes on the Great Pyramid of Egypt.[1]

The mode of inquiry adopted there, produced, within the limits of errors of measurement, such satisfactory answers touching the mental ideas and intentions of the long-departed designer and builders of that ancient structure,—that a desire was naturally felt to try how far the same principle could be carried, as a general method of investigation, wherever any remains of very high antiquity still exist.

From many such attempts, during twelve months past, the materials have been obtained for the present essay upon the *Antiquity of Intellectual Man*. An essay of no very great length,

[1] *Life and Work at the Great Pyramid of Egypt, during the Months of January, February, March, and April,* A.D. 1865, *with a Discussion of the Facts ascertained;* by C. Piazzi Smyth. Edinburgh: Edmonston and Douglas, 1867.

and admitting doubtless of much future improvement as well as extension, but based on a more direct series of solid, trustworthy, and contemporary ancient facts, than have ever yet been utilized on a regular system for this special purpose; viz., to approximate towards real human history, whatever that may have been, at the most distant periods of time.

As those epochs embrace dates at upwards of 1000 and even 2000 years before the birth of all school science; *i.e.,* before man began, either amongst the priests of Egypt or the poets and philosophers of Greece, to endeavour to acquire a knowledge of nature by his own observations and measures,—they inevitably belong to those very early ages of the world, concerning which it is asserted just now, in too many quarters, that the Sacred Scripture says one thing, and modern science the very opposite.

At this particular stage then, in the progress of education and learning,—and when the old arguments on either side of the controversy appear to have been abundantly displayed,—a new method of investigating some of the same questions has appeared, and is employed in the following pages. A method sufficiently scientific, because based chiefly on practical astronomy ; and abundantly direct, because dealing only with the very contemporary remains of the times actually concerned.

What this method therefore says,—within such limits as it has the opportunity of testifying,—not only cannot be insignificant to any man now living, but will probably be found to throw a new and explanatory light upon several passages in Genesis, usually considered as serious difficulties.

The author would have had far less confidence in submitting the results of his labours in so rich and considerable a field, to the public thus soon,—but that, while two items of these labours have had the misfortune to excite personal feelings which he sincerely regrets, and even to entail a vindictive opposition carried on before the Royal Society of Edinburgh and elsewhere during the past winter,—there has resulted thence a most signal advantage to a new theory ;—viz., that of being speedily and thoroughly tested by the able heads and very unsparing hands, of, as they say themselves, many opposing mathematicians, arithmeticians, doctors, and engineers, both military and civil; and every assailable point, even down to an unimportant printing error, has, it may be hoped and expected, been discovered ere this, and made the most of.

So material has the author considered these involuntary co-operations with him in revising his pages, correcting his language, and freeing his calculations from stray errors,—that he has laboured everywhere to give to such critiques almost as full space and representation as to his own views of the same cases. Where the objectors have gloried, and still glory, in their remarks, and have printed them publicly, considering them apparently as triumphant representations of their side of the question,—the author has not hesitated to attach their names to their writings, and this whether he agreed or whether he disagreed with them; but where the critics have been more doubtful about the correctness of their opinions, —and he has himself been able subsequently to prove them in error,—he has kept back the name, but employed inverted commas on the words used, in order to indicate that such opinions *are* actually floating about in society at present.

There will thus be found, the author trusts, a fulness and fairness about every part of the book, —which will be approved as much by those holding opposite opinions to, as those thinking with, himself; while the reality and directness of its style and results, are only a necessary consequence of the practical method here employed, for investigating a subject usually rather hazy when treated of from the literary side only. Many points, as the author is only too well aware,—and points quite proper to be discussed according to the method of the book,—are touched on too shortly, and perhaps superficially, by him. But some of these topics have been fortunately taken up in a series of Appendices (pp. 393-502), contributed chiefly by a few very earnest inquirers into, and able prosecutors of, that paramount problem, *the antiquity of intellectual man;* and their thoroughly-searching and conscientious labours in the cause, it is to be hoped, will not soon cease.

CONTENTS.

CHAPTER 5.

CHAPTER 6.

CHAPTER 7.

CHAPTER 8.

CHAPTER 9.

CHAPTER 10.

CHAPTER 11.

CHAPTER 12.

CHAPTER 13.

CHAPTER 14.

CHAPTER 15.

ARCHÆOLOGICAL DATA.

CHAPTER 16.

CHAPTER 17.

CHAPTER 18.

CHAPTER 19.

CHAPTER 20.

CHAPTER 21.

INTELLECTUALITY OF THE G. PYRAMID.

CHAPTER 22.

CHAPTER 23.

CHAPTER 24.

CHAPTER 25.

CHAPTER 26.

CHAPTER 27.

CHAPTER 28.

Units and Standards of Great Pyramid Measure.

CHAPTER 29.

CHAPTER 30.

CHAPTER 31.

CHAPTER 32.

CHAPTER 33.

CHAPTER 34.

CHAPTER 35.

APPENDIX 1.

APPENDIX 2.

APPENDIX 3.

APPENDIX 4.

APPENDIX 5.

APPENDIX 6.

APPENDIX 7.

APPENDIX 8.

APPENDIX 9.

EXPLANATION OF PLATE.

The Plate which forms the Frontispiece is intended to show the rise and progress of ancient architecture in many various countries, and through all times extending from 2500 B.C. to 1500 A.D. To this end, the first column on the left hand, contains a scale of dates, beginning with the earliest at the top, and ending with the latest at the bottom; and each date marked in this column is supposed to extend transversely or horizontally across the whole table.

In the successive vertical parallel columns extending towards the right, from the date column, the amount of architecture existing in the world is shown by the amount of black markings in the said columns. The date of such architecture is ascertained by reading off its level on the scale of dates in the left-hand column; and its geographical position by looking to the title at the head of the column where such black marking may occur.

Example 1.

In the 4th column (first half) of 'Localities,' a maximum breadth of black marking appears very decidedly at one point. The level of that point on the left-hand Time scale is '1000 B.C.;' while the name at the top of its own part of the column is 'Palestine.' Hence the maximum indicated, refers to the buildings carried on by King Solomon.

Example 2.

In the 7th Locality column, under head of North India, the higher part of the black marking is in thin dots only.

That is to indicate that the traces of ancient architecture there, rest only on tradition, long subsequent literary records, and other uncertain data. Where the line of dots swells out into a black line or band,—there are absolute, material, and still existing proofs to refer to.

EXAMPLE 3.

In the 2d Locality column, under head 5, or Assyria, the black band which begins at the top gradually from a point, near 2000 B.C,—ends suddenly near 600 B.C., in order to denote the total destruction of Nineveh, and the termination of the Assyrian empire at that time.

EXAMPLE 4.

In the 1st Locality column, under 1, or 'Great Pyramid and Memphis,'—the black marking begins at once very broadly, and decreases as it descends, through long periods of time. That is to indicate the suddenness with which the Great Pyramid appears on the scene of the world's history; the highest of all buildings ever erected, even up to the present time,—one of the best constructed too; and without any certainly known or worthy predecessors in stone architecture; while its successors gradually dwindled down both in size and quality of composition.

MENE, MENE, TEKEL, UPHARSIN,

NUMBER, NUMBER, WEIGHT, DIVISION.

DANIEL v. 25.

CHAPTER 1.

OF THE *CONTEMPORARY* METHOD OF INQUIRY, INTO THE ANTIQUITY OF INTELLECTUAL MAN.

NOT seldom has it been remarked, that the longer man lives upon the earth, and the more the ground grows ancient beneath his feet,—so much the more does he desire to become acquainted, if it be possible, with the history of the first of his species.

Nor have his earnest desires in this direction been altogether ungratified; for, notwithstanding that the immense abyss of time which separates any existing, from the primitive, race of mankind, is perpetually widening,—yet the fact is certain, that much more is known now, and coming to be known year after year, concerning the early days of the earth and its first intelligent inhabitants,—than during many centuries and even tens of centuries that have preceded us.

This successful example of hoping against hope,

and succeeding to overcome even apparent impossibilities, is with little doubt due to the general advance of modern science, to the adoption of sure methods of research, safe principles of philosophizing, and especially to a more accurate knowledge of nature, 'in number, weight, and 'measure.'

To such things certainly is more particularly due that increased call—which has been so extensively experienced of late—for renewed examinations into old traditions and even long established beliefs. While we all know, that such examinations have not seldom resulted in freeing the necessary history of the world, from a vast amount of classical fables, and legends of romance, very difficult to believe in. They seem to have established also, the generally sound principle,—that no man is much to be trusted when he is writing about any other than, either his own times, or those of which he has *contemporary* information. Or again, as some eminent Continental philosophers have phrased it,—'the history of no 'country can be securely carried back, much 'beyond the date of its earliest *contemporary* 'monuments.'

This is the expressed opinion of, amongst others, the erudite Dr. R. Lepsius; and no one will venture to accuse him of want of scholastic power to appreciate the letter, or spirit, of any classical author who has written on primitive times: while his more impetuous friend, the late Baron Bunsen, after wading deeply through many such old historians, exclaimed bluntly—'but now it is 'high time that we should leave all this *chaff*, 'and turn to the monuments;' *i.e.*, to things contemporary in their origin with the ancient events he was investigating.

CHAPTER 2.

BUILDINGS ALMOST THE SOLE AVAILABLE *CONTEMPORARY* MONUMENTS.

Now, when we look around for any such contemporary remains of the earliest possible ages,—it is not books or rolls of writing, but something more or less in the shape of *buildings,* which answer to our call.

No doubt Hindoos may boast of the 20,000 years of age of some of their sacred hymns, and Egyptians of the 30,000 years of their books of Hermes; but they cannot show *contemporary* copies either of that time or of its mere twentieth part. Yet are there buildings in the world, still standing open to examination, which were erected several thousand years before any known copy of any of those books was transcribed.

Ancient buildings, then, if they still exist at all, have both antiquity and contemporaneousness extreme, on their side. But, seeing that our title restricts us to the antiquity of *intellectual* man,—do they (these ancient buildings) possess also legible traces of mind, and proofs of true intellect?

Several writers, whose studies have been much in that direction, seem now inclined to answer affirmatively; at least for all the oldest instances. And one of these very advancing students, an architect by profession, and personally acquainted with the grandest expressions of his art, throughout three-fourths of the habitable globe,—has particularized in forcible language why and wherefore this most important science of ancient buildings, or to him *archæology*—has not been allowed hitherto all that importance to which he thinks 'it is undoubtedly entitled.'

'We live,' says he, 'in an age when all art (at least build-'ing art) is a chaos of copying and confusion; we are daily 'masquerading in the costume of every nation of the earth, 'ancient and modern, and are unable to realize that these 'dresses in which we deck ourselves were once realities. 'Because architecture, since the Reformation in the sixteenth 'century, has in Europe been a mere *hortus siccus* of dried 'specimens of the arts of all countries and all ages, we cannot 'feel that, before that time, art was earnest and progres-'sive; and that men did what they felt to be most appropri-

'ate, by the same process by which nature works. We do 'not therefore perceive that, though in an infinitely lower 'grade, we may reason of the works of man before a given 'date, with the same certainty with which we can reason of 'those of nature.—When this great fact is once recognised— 'and it is indisputable—Archæology and Palæontology take 'their places side by side, as the guiding and vivifying ele- 'ments in the sister sciences of Ethnology and Geology; 'and give to each of these a value they could never otherwise 'attain.'[1]

The same writer then explains more precisely, thus—

'Almost all men who have hitherto written on these '(primeval history) subjects, have derived their information 'from Greek and Roman texts; but, if I am not very much 'mistaken, these do not suffice. The classic authors were 'very imperfectly informed as to the history of the nations 'who preceded or surrounded them; they knew very little of 'the archæology of their own countries, and less of their 'ethnography. So long, therefore, as our researches are 'confined to what they had written, many important problems 'remain unsolved, and must ever remain as unsolvable as 'they have hitherto proved.'

As a remedy therefore for so hopeless a state of things, that writer urges what he calls 'the '*lithic*,' or the early contemporary *building* method of investigation; and protests of it that—

[1] *A History of Architecture in all Countries, from the Earliest Times to the Present Day.* By James Fergusson, Esq., F.R.S. London, 1865.

'When properly studied and understood, there is no language so clear, or whose testimony is so undoubted, as that of those petrified thoughts and feelings which early men have left engraved on the walls of their temples, or buried with them in the chambers of their tombs. Unconsciously expressed, but imperishably written, they are there to this hour. And no one who can translate them, can for one moment doubt, but that they are the best, and frequently the only, records that remain of bygone races.'

With these very elevated views of his profession, Mr. Fergusson has examined almost all the existing remains of architecture, and with a success and discrimination which leave very little indeed to be added by any one else. In fact, hardly anything more could have been desired, had that excellent artistic and thoroughly professional architectural writer, only been, on certain occasions, more widely scientific.

Some attempt, therefore, to supply this omission at one point—and only for the beginning of that vast stock of important data belonging to many ages, which, both his admirable industry has collected together, and his method and acumen have marshalled into a clearness of order never presented before—is all that I now propose to venture upon ;—the first collateral branch of inquiry referred to, being the astronomical.

CHAPTER 3.

ASTRONOMY, THE FIRST PROBLEM TO BE GRAPPLED WITH, IN ANCIENT BUILDINGS?

IN the early part of last year, a statement was put forward very confidently [1] in one of the larger class of quarterly reviews, to the effect that—

'Pyramidal piles of earth and stone are the peculiar marks 'by which we may discover the sites of the earliest settle-'ments of mankind. The idea of such piles first appeared in 'the valley of the Euphrates, and culminated in the valley of 'the Nile.'

This is remarkable enough as an assertion, and most extensive in its range. But the author adds still further, and more remarkably—

'Whatever their forms, or wherever situated, in Asia or 'Africa, *one condition* is common to them all; intended 'primarily for astronomical observatories, the sides of each 'accurately correspond with the cardinal points.'

---

[1] *The Edinburgh* (by name, though published in London) for April 1867; in an article upon 'American Antiquities.'

The intention implied above, may well be put on one side just now, while we examine into the *condition* asserted to exist; for that is decidedly the first point to be proved, before anything else can be entered upon safely. And such *condition* —as to the correct astronomical direction of the walls or foundations of a building—is precisely one of the facts which can be determined much more easily and accurately in the present day, than it could be, or have been, in the times of Herodotus, Strabo, Diodorus Siculus, or their many followers.

Without therefore at this time subscribing in any degree to the opinions of the learned essayist, —let us carry his asserted astronomical law, or condition, of ancient buildings along with us; and use it as a proposition to be tested,—while we glance over a general arrangement and classification of the several structures themselves; taking first, whatever may be more or less strictly included within the domain of *Architecture.*

ARCHITECTURAL DATA.

CHAPTER 4.

VALLEY OF THE NILE.

Lower Egyptian Pyramids.

AFTER travelling over the whole extent of his important and varied subject, and even attempting by its means some special excursions into the characteristic differences of primitive races and supposed varieties of mankind, Mr. Fergusson gives it as the settled result, of all impartial and practical investigation from his professional point of view, that,—do what we like, or 'stretch the 'history of architecture how we will,—we cannot 'get beyond the epoch of the Pyramid builders' (of Lower Egypt).

With those Pyramids, therefore, as the earliest remaining buildings of the whole earth,[1] he begins

1 'The earliest known architecture, the Pyramids of the 4th 'Dynasty, exhibit simple forms of vast magnitude, and of the 'minutest finish.'—*Synopsis of British Museum*, 60th edition.

his most philosophic and vividly ethnographic history of architecture. Nor does he stand alone in such a view; for, with *all* the great Egyptological scholars of modern times,—Lepsius, Bunsen, Gardner Wilkinson, Osburn, Mariette Bey, Renan, Rawlinson, Lesueur, and many others, —Egypt is held to be the most ancient civilized country with monuments still existing. 'Egypt,' says Baron Bunsen, 'is the monumental 'land of the earth, as the Egyptians are the 'monumental people of history.' The first and the earliest of all architectural nations, and living in a climate superior to that of any other country for preserving material fabrics once erected, its monuments are, therefore, remarkable storehouses of *contemporary* records of early time; and, of all those monuments, the Pyramids represent Egypt's most ancient days of human life and toil.

The *absolute* dates for those very early times, as given by the several Egyptological authors, are indeed often at variance amongst themselves; but they agree well in their *relative* chronologies, and assign, therefore, as will appear in the sequel, to all other practically known buildings in Egypt, and even in every other part of the world, a lower date than they attribute to the chief of those wonderful *pyramidal* remains, which are still found in remarkable preservation on the Lower Nile Valley. On a very small part of it too; for they are all confined to its western desert flank, and within the latitudes of 30° 3′, and 29° 17′, north.

How then do these most ancient of all existing buildings,—and which we have here in an isolated and easily accessible group,—correspond to the proposed astronomical law?

A simple question; and yet, before entering upon its answer, it may not be inexpedient to set in order our ideas, touching both the pervading shape, and general appearance, of the buildings concerned.

Figure of the Pyramids.

Of all mathematical solid figures, few are so easily defined as 'a Pyramid;' while, of all human buildings, none have ever been erected so extremely near to being exact representations of a pure geometrical idea, and of nothing else, as 'the Pyramids of Lower Egypt.'

But there are varieties among the Pyramids of

pure mathematics; and all the Egyptian practical examples represent one of these varieties only, viz., that possessing a square base. Equally would they have been 'Pyramids,' had their bases possessed three, five, six, or any other number of sides; but four was chosen, and is repeated unerringly in every known case.

The sloping faces or flanks of such a Pyramid, will next be equal in number to the sides or edges of its base; and in form, will be flat or plane triangles, fastened broadly on to the base-sides below, and meeting together, with their upper acute angles, or vertices, in a point above. These triangular faces may be of any degree of acuteness or obtuseness towards their summit, according to the desired height of the Pyramid, proportioned to its breadth of base; but every such sloping flank in one Pyramid, must be closely similar[1] to its fellows; just as it is also necessary, that the lengths of the several sides of the base should be equal, and the said base decently flat.

Theory further requires, that a Pyramid shall be a *solid* figure; and the early Egyptian archi-

[1] Theory, indeed, insists on absolute exactness, but practice must be content when the error of construction does not amount to more than a small proportion of the whole.

tects so far complied with the demand, that, although some of their Pyramids are the largest buildings in the world, their only hollow internal spaces are two or three small rooms; and certain narrow passages of access,—so low that they can only be entered by men in a stooping posture.

These very limited hollow spaces, do not form much more than one thousandth part of the whole mass,[1] all the rest of which is built solid with stones regularly squared and laid in cement; or with bricks, and then usually cased in stone,—stone, too, of a very fine order; and,—except when partially of red granite,—white, compact, admirably worked, true at the joints, and planed down or ground flat on the surface, with astonishing success.

This, then, was the pristine appearance of the earlier Egyptian Pyramids; viz., square-based below, pointed above, and triangularly flat-sided; fair and bright to behold, like a snow-capped mountain peak shaped in a crystalline mould.

Time, however, has worked many changes with them; lightning has split the tops of some;

[1] In all ordinary architecture, both ancient and modern,—everything large, in fact, but the Pyramids,—the hollow, vastly exceed in bulk the solid, parts; the latter only forming, especially in recent times, a thin and too perishable shell.

earthquakes at one age,[1] and at another the destroying hand of man, during many laborious years, have pulled off the fine casing from others; and those built originally of brick, though by no means the oldest of the series, have fallen down into round-topped mouldering heaps. But in all these instances, the very ruins of the upper part have been the direct means of preserving both the bases and the beginning of the sloping, and triangular, sides; so that a moderate amount of excavation now, often reveals at once the original fiducial points of commencement; and enables us, besides recovering the primeval *figure* of the monument as above, to study its astronomical emplacement,—our question of page 8, so long delayed.

Astronomical Emplacement.

Already it will be seen that the characteristic, *four* sides of the Lower Egyptian Pyramids' bases make them suitable to being tested by, or upon,

[1] In the year 908 A.D., seventy-eight years after Al Mamoon's forcible entrance into the Great Pyramid, Egypt was visited with so severe 'a tempest and earthquake,' as to form the chief historical event of the time, and not improbably to have assisted in bringing down the outside casing-stones from many of the Pyramids.

See also *Life and Work* by the author, vol. ii. p. 103.

the four cardinal points. And if it now again be inquired, how they suit the asserted law of these sides being directed accurately on those natural and geographical points or directions, the first general answer is, 'remarkably well.' For, with a single exception, all these Pyramids are described and stated, by nearly every scientific traveller who has visited them within the present century, to be as truly oriented (as to their sides, upon the cardinal points) as the respective modern engravers employed by these travellers have been able to represent them on paper.

This illustration, indeed, must not be taken in the present day as a proof of perfect accuracy, though it indicates a probability of the errors being small. But in the case of the Great Pyramid, which for other reasons is generally considered almost, if not absolutely, the oldest of all,—I can say from my own express astronomical observations upon it, that the errors of its eastern and northern, if not also its western and southern, base-sides, are under 5 minutes of angle;[1] a smaller quantity than most engravers'

[1] See *Life and Work at the Great Pyramid*, vol. i. p. 548; vol. ii. p. 196; and vol. iii. p. 107.

plans are capable of showing. Less too is this, than one-third part of the error committed in the astronomical emplacement of the most celebrated European observatory of the latter part of the middle ages; viz., the palatial structure of Tycho Brahe at Uranibourg; and not intentionally approached by any other known building, through 3700 years downwards, from the Great Pyramid's date of foundation.

Now the rest of the Pyramids, not having been examined with the same accuracy, neither as to seeking out the original fiducial markings, and corner-sockets, of the builders, nor with the same power of astronomical instrumentalism,—we cannot assign to them, unquestioned, the same full amount of exceeding excellence in a difficult scientific feature. From some partial measurements, indeed, of my own, on the second Pyramid, we may safely infer that it is not *far* from as accurate as the Great Pyramid; while we may not improbably, but in a more distant degree, conclude that the third and other smaller Pyramids, though evidently built with less pains and skill, and more attention to the ornament of colour and costly material, than the Great Pyramid,—are yet of the same school; followers, though in a most humble degree, of their mighty, but plain and scientific prototype.

All the above Pyramids, are those which form the well-known group of Jeezeh, in a mean latitude of 29° 59′. But if we visit the other groups further south, although there is still a family resemblance,—all square-based Pyramids, and all set with their sides approximately directed to the cardinal points,—yet their variations from the angular slope of the Great Pyramid, and from its materials of construction,—go on continually increasing; while one, at least, of their number, the so-called Great Pyramid of Sakkara[1] is reported by Howard Vyse and Mr. Perring to be no less than 4° 35′[2] from the cardinal-point direction.

We need not suppose that any others are quite so far from the truth of nature and the example of the Great Pyramid, as this one. But the

[1] This Pyramid, for some time thought, on account of its ruinous condition, to be the oldest of all the Pyramids, has more recently been shown to have acquired that character from bad construction and worse material; and to be considerably more recent than the Great Pyramid, or that of Jeezeh; as Dr. Lepsius accordingly places it, in the folio volumes of his *Denkmaeler*.

[2] By a pure misprint, the quantity is given in Mr. Fergusson's plate, p. 90, as only 4′ 35″.

Sakkara case is nevertheless a luminous example of the growing neglect with which following men treated the higher scientific conditions. For, while Dr. Lepsius is very confident in placing the Great Pyramid as the oldest monument of Egypt,[1] and William Osburn successfully shows that from that Jeezeh hill point, or the crown of the Delta as a beginning, the early progress of Egyptian agriculture and architecture *ascended* the banks of the Nile,—we may, in an approximate manner, conclude every Pyramid more recent, in proportion as it lies further south; and then we have the following tabular view of never-ending deviations:[2]—

[1] See p. 21 of Dr. Lepsius's *Letters from Egypt in* 1842, *and* 1845. See also *Life and Work at the Great Pyramid*, vol. iii. p. 364.

[2] This table is composed almost entirely from the data procured by Colonel Howard Vyse, and Mr. Perring; and published by the former in his celebrated volumes on the Pyramids of Jeezeh.

His first so-called Pyramid, or that of Abou Roash, in lat. N. 30° 3′, is indeed here omitted. Partly, because he himself states, 'that no part of the external casing is to be found; indeed, the 'edifice was not probably ever completed, or even raised to a con'siderable height; for scarcely any materials, and very little rubbish 'are to be seen, although the situation is very difficult of access.' And partly, because his attempted identification of the site, with that of the Pyramid-famed Cochome of Ancient Egypt, has been since displaced by Lepsius and Osburn in favour of Sakkara.

PYRAMIDS OF LOWER EGYPT, APPROXIMATELY.

NAME OF PYRAMID.	Latitude North.	Angle rise of faces with horizon.	Length of one side of base (always 4 sided).	Deviation of sides of base from cardinal points.
			British Inches	
Great Pyramid of Jeezeh	29° 59′	51° 51′	9142	0° 4′ 35″
Second of do. . .	29 59	52 20	8493	Supposed to be small.
Third „ . .	29 58	51 0	4254	Do.
Fourth „ . .	29 58	in steps	1230	Do.
Fifth „ . .	29 58	52° 15′	1749	Do.
Sixth „ . .	29 58	in steps	1230	Do.
Seventh „ . .	29 59	52° 10′	2070	Do.
Eighth „ . .	29 59	52 10	2070	Do.
Ninth „ . .	29 59	52 10	1920	Do.
Zouyet el Arrian, .	29 57	ruined	3600	Do.
Pyramid of Reegah (with a granite basement at angle of 75° 20′), . . .	29 56	52° 0′	1480	Do.
Northern Pyramid of Aboosеer, . . .	29 54	51 43	3084	Do.
Middle Pyramid from Aboosеer, . . .	29 54	51 (?)	3288	Do.
Great Pyramid, Aboosеer, . . .	29 54	52 (?)	4312	Do.
Pyramid 2 at Sakkara,	29 53	52 (?)	4775	Do.
Great Pyramid of Sakkara (the base of this Pyramid is larger from E. to W. than from N. to S.	29 53	73° 30′ (in steps)	3712 / 4727	4° 35′ 0″
Pyramids 4, 5, 6, 7, 8, and 9, at Sakkara,	29 53	ruinous	3000 (?)	Supposed small.
Mastabat el Faraoon,	29 53	steps (?)	3708	Do.
North Brick Pyramid of Dashoor, . .	29 49	51° 20′	4200	Do.
North Stone Pyramid of Dashoor, . .	29 49	43 36	8633	Do.
South Stone Pyramid of Dashoor; the upper half flatter than lower, . . .	29 48	54 5 / 42 59	7400	Do.
South Brick Pyramid of Dashoor, . .	29 48	57 20	4110	Do.
Pyramids of Lisht, .	29 38	ruinous	5000 (?)	Do.
Pyramid of Meydoom,	29 27	74° 10′	(?)	Do.
Pyramid of Illahoon,	29 17	(in cliffs ruinous)	4320	Do.

Variations with Time.

To the particulars given above we might add, that almost from the first, or the Great Pyramid itself, a desire to introduce ornaments, foreign to the mathematical theory, appears; so that the second Pyramid has a thin red-granite basement-course, and the third Pyramid has a similar richly coloured granite casing, mounting a long way up its slopes.[1] On the east side of the second Pyramid, too, but much removed from it, stands a peculiar kind of temple. On the same side of the third Pyramid stands something of a similar kind, better preserved, and not so far off. And with some of the southern Pyramids, similar hollow temples are brought much closer up to their respective Pyramids—perhaps into contact, and on their north as well as east sides, and have more or less painted and decorated interiors.

While again the Great Pyramid enters only at the north side, and that flank is constituted gene-

[1] The original third Pyramid, as I believe, first shown by Baron Bunsen upon Colonel H. Vyse's and Mr. Perring's drawings, was enlarged subsequently, and, according to Dr. Lepsius, very long subsequently, to its first building. Whatever, therefore, we now see outside, belongs to this later age.

rally the entrance side of all the Pyramids, there are some later ones which enter from two or more sides, and have very conflicting suites of apartments within. Used, indeed, they always were, more or less as tombs, and probably royal ones: but not according to the principles originally set forth in the first of these most peculiar geometrical buildings.

Hence then this most remarkable line of *square-based*, but symmetrically triangular-sided, Pyramids of Lower Egypt, after having begun most successfully with an example, wherein (as we shall hope to show presently) both its shape, size, and astronomical emplacement include important data of science, pure as well as applied,—this line of imitations fell wofully away afterwards, from all such high and characteristic ends and aims. Indeed, it would appear by the enduring facts of their successive constructions, that the ancient Egyptians must have writhed under and resisted, much more than have willingly and spontaneously cultivated, the idea of their primitive most mathematical Great Pyramid; wherever that idea was originally obtained from, or whatever purpose it was first invented for and ordered to be constructed to subserve, on so enormous a scale, as well as in so enduring a manner.

Occasionally, as the facts indicate, the people worked themselves up into some amount of enthusiasm, or were flogged into harder labour by a more than usually despotic king, so as to build rather a bigger Pyramid for the said monarch's future more glorious entombment, than their last architectural performances. But these were mere dry revivals, which brought out no new thoughts, while allowing many of the old ones to perish. After a few such spasmodic flashes, therefore, at separate, distant intervals, the light of practical Pyramid-building, as any vital influence, went out for ever to that land and that people. Their Pyramids had become, through several centuries, continually smaller and less perfect, even constructively and mechanically; while they had entirely lost every trace of science symbolized for distant posterity. Then the sons of Mizraim left off building Pyramids altogether, and turned their attention to an entirely different style of architecture. Nor did they ever return to the Pyramid idea, though they had the most splendid period of their empire, to be, in wealth, extent of population, and showy architecture, still before them.

Dates of Egyptian Pyramid-Building.

The period, then, during which this strange light of Pyramid-building shone on Egypt, was limited, and rather sharply defined in the history of that country.

Beginning with a most grand example,—or perhaps with one or two imperfect attempts preliminary to that triumph of art and early science 'the Great Pyramid,' (though that will be a point for elucidation in the subsequent portion of our book); and then, after a period during which every succeeding monarch imitated his predecessor in building a Pyramid,—sometimes lazily, sometimes energetically, never wisely,—after such a period, pretty active for a time, but always with a downward tendency,—then came the last puny, despised example; after which Egyptians washed their hands of the art, and acted as though they would have the world believe, that through all their history they never had had anything to do with the, to them, for some unknown reason, accursed thing of Pyramid-building.

The date of both that most noteworthy beginning and peculiar ending in Coptic architecture, Mr. Fergusson would place between 3900 B.C. and 2600 B.C.; Lesueur, Renan, etc., between 5400 B.C. and 4000 B.C.; Lepsius, nearly as Fergusson; Gardner Wilkinson and Rawlinson, between 2500 B.C. and 2200 B.C. But William Osburn[1] places it between 2300 B.C. and 1900 B.C.; and as his dates are sustained by astronomical deductions begun by Sir John Herschel and continued by myself,[2]—from what may truly be called, in Mr. Fergusson's own words, 'the very petrified 'thoughts, feelings, and intentions of the Great 'Pyramid builders' when employed thereon, as a *contemporary* monument,—of course I shall adhere to Mr. Osburn's numbers; which are besides, the result of a very rigid and thorough examination of all specially hieroglyphic and Egyptologic data.

Two remarks further, however, may be advisable here, before descending the stream of time to see what the Egyptians took to building, *after* they deserted the Pyramids.

[1] *Monumental History of Egypt*, vol. i. Gardner Wilkinson also, in his earlier Works.

[2] *Life and Work at the Great Pyramid*, vol. iii. pp. 215-299.

The *first* such remark, is—seeing that the enormous differences of dates among the Egyptologists are chiefly due to some sort of stretching out, more or less, of the very same number and order of positive events, so as to occupy more, or fewer, years; and as this Procrustean method prevails increasingly during the earlier part of Mizraite history,—we must never take a date from one of those modern Egyptological gentlemen, and put it down in the chronology of another of them, without first reducing the former's system to the value or terms of the latter. For, without such a necessary reduction, the whole facts of kings and reigns would presently be displaced *relatively*; and posterity might be set above their ancestors, with the most disastrous facility.

We shall close this chapter, therefore, with a tabular view of the first twenty Egyptian dynasties, as dated by the several Egyptological authorities; hoping to supply therein a simple inspection method of making the correction just mentioned. And now comes our *second* word of advice.

Though there are at present such large discrepancies amongst the Egyptologists, they follow some sort of law; and there *is* a science binding them to a certain extent together, and making them a very respectable body of learned men. Philosophers they are, who will no doubt go working on and on, until they have brought their mutual differences within a smaller compass, and have much improved the whole world's knowledge of their most peculiar subject. But on or near their confines are to be found stray geniuses, erratic stars, meteors blazing up here and there anomalously: in fact, Egyptological analogues of anti-Newtonian gravitation men, and seekers after perpetual motion. Of and against all such, let the earnest student be forewarned and forearmed.

One of these solitaries, who has written much on the history of Egypt, and, when touching on her periods of Roman bondage, very socially and pleasantly,—actually makes King Suphis of the 4th dynasty and the Great Pyramid, go forth from his home in Memphis, to fight against kings of the 12th dynasty, in Thebes; and this about the year 1650 B.C.; that is, at a date long after the widely removed lives, and far separated deaths, of both elder and younger monarchs; or when they could certainly do no harm either to each other or to their respective kingdoms.

Another of these disturbers of history is, however, a man of higher mark, being no other than the late Sir George Cornewall Lewis; an excellent classic, but more Greek than all the Greeks in this one point of their national weakness, viz., thinking all the rest of mankind illiterate barbarians, and deeming nothing older than, or half so good in the world as, themselves. Hating therefore, and despising also, the Egyptians in his Grecian heart, knowing nothing of their practical constructions or hieroglyphics, but deriving some slight account of them from the Greeks only, and they writing merely for their national purpose of exalting themselves,—Sir George Cornewall Lewis rather comically informs his readers, that the first Greek Olympiad, or 776 B.C., is the earliest date on which we can rely for any event in the past annals of mankind.

For the Egyptian Pyramids, indeed, he does go a little farther back; but thinks that they cannot be older than 1012 B.C. How, therefore, he stands among the modern hardworking Egyptologists, who gather their knowledge of Egypt by applying their own brains to Egyptian monuments,—the following table will show; *i.e.*, as touching the usual 'dynasties;' for, though some writers begin their Egyptian histories long before even the first Manethoan reign proper, we shall discuss *them* in a separate chapter.

EGYPTIAN CHRONOLOGY, showing the Date of the Beginning of each Dynasty *approximately*, according to various authorities.

Number of Dynasty.	Lesueur, Mariette Bey, Renan, etc.	Lepsius, Bunsen, Fergusson, etc.	Lane, Gardner Wilkinson, Rawlinson, etc.	William Osburn and Astronomical Calculation.	Prevailing Architecture.
	B.C.	B.C.	B.C.	B.C.	
1	5735	3892	2700	2429	Pyramid Builders.
2	5472	3639	2480	2420	
3	5170	3338	2670	2329	
4	4956	3124	2440	2228	
5	4472	2840	2440	2228	
6	...	2744	2200	2107	
7	...	2592	1800	...	
8	...	2522	1800	...	
9	...	2674	2200	2107	
10	...	2565	...	1959	
11	...	2423	2200	2107	
12	3435	2380	2080	...	
13	...	2136	1920	...	Palace-Temple Builders.
14	...	2167	2080	...	
15	...	2101	2080	1900	
16	...	1842	1800	1900	
17	...	1684	1776	...	
18	...	1591	1520	1674	
19	1314	1443	1324	1394	
20	...	1269	1232	1314	

ARCHITECTURAL DATA—*continued.*

CHAPTER 5.

VALLEY OF THE NILE—*continued.*

Theban Temples.

A BLANK follows in the history of Egypt after the last degenerate specimens of the Pyramids had been erected,—a long stretch of Nile banks nearly devoid of architectural remains is passed, as a modern traveller—all unconsciously pursuing the path of early Egyptian empire—ascends the river towards the south: and then, in the one case, we arrive at the Theban kingdom, flourishing in its chiefest glory from 1800 to 1200 B.C. according to most of the chronologies, here not very different from each other; and in the other case, we practically reach the plain of Thebes in Upper Egypt, under the parallel of 25° 40′ N. latitude; with the effect of finding it all gorgeous with temples and palaces.

Grandly shine those ancient[1] glories of architecture in sandstone and limestone, granite and porphyry, basalt and alabaster. But there are no Pyramids amongst them.

Some archæologists, catching at the remembrance of a form which had set for ever,—will persist in calling attention to the slightly pyramidal, or inclined, faces of many of the public buildings, and to the outer surfaces of Theban door-posts and 'pylon' towers. These last being strong, yet decorated, hollow structures near the entrance of every temple, and forming the glory of Thebes in its days of renown: for it was then the 'city of a hundred temple pylons,' not city *gates*, as Homer and his translators erroneously phrased it.[2] But the slope of these buildings, is not even approximately the slope of the Great Pyramid; while instead of an almost solid fabric (with $\frac{1}{1600}$ only of hollow), they have merely *apparent* masses, which are *one-half* or more of hollow space within, or only positive open colonnade.

[1] Not ancient as to the Pyramids of Lower Egypt, though extremely so as compared with the temples of Greece and Rome.

[2] Thebes had no surrounding walls, and therefore no city gates. Its temples were its strongholds, and were separate and isolated holds; the houses of the citizens being scattered loosely abroad.

In place too of smooth, flat, simple walls, those palace temples, or temple-palaces,—for no one seems to know exactly which,—show obelisks, pillars, capitals, cornices, and pediments, sculpture in various degrees of relievo, as well as the round; with multitudinous engraved adornments of every kind, and all of them painted with vivid pigments up to the very eyes. In place also of silence and purity as of old, they are covered (these Theban temples) from end to end, and from top to bottom, with enormous hieroglyphic inscriptions, bearing the praises of their despotic and murderous kings, such as Thothmes and Amenophis and Rhamses Sesostris; and are polluted with all the abominations of the religious worship of a thousand animal-headed gods, each one more vile than another.

This is the ordinary '*ancient Egypt*' of most of our historiographers. It is that also, whose remains both our travellers so rave to see, and our collectors delight to bring home pretty specimens of the idols of,—from little green images the size of a fly, up to the monster granite colossi in the British Museum.

The Astronomical Test.

In the midst of so much falling away, then, into their own deceitful religious inventions, how fared the asserted astronomical *condition*, touching the orientation of the walls of these showy buildings by Egyptians of the second empire?

Mr. Fergusson says, that 'they were placed 'anywhere, facing in any direction, and generally 'affected with a *symmetriphobia* that it is difficult 'to understand. The pylons are seldom in the 'axes of the temples; the courts seldom square; 'the angles frequently not right angles, and one 'court succeeding another without the least re-'ference to symmetry.'[1]

Checking this account by reference to the plates in Dr. Lepsius's large work on Egypt,[2] a most ample confirmation is found there; for, if

[1] And again, at page 109 of his *Architectural History*:—'The 'palace of Luxor is further remarkable as a striking instance of how 'regardless the Egyptians (of the Theban age of kingdom) were of 'regularity and symmetry in their plans. Not only is there a con-'siderable angle in the direction of the axis of the building, but the 'angles of the court-yards are, in scarcely any instance, right angles. 'The pillars are variously spaced, and pains seem to have been gra-'tuitously taken to make it as irregular as possible in nearly every 'respect.'

[2] *Denkmaeler aus Ægypten un Æthiopen*; 12 folio vols.; Berlin.

the direction of the look-out of each temple be taken, as defined by the longer general axis of the whole, passing out from the inner penetralia and proceeding forth through the pylon towers, the following numerical results are obtained for their astronomical directions :—

ON WEST SIDE OF NILE.

Northernmost temple of Sethos I., .	=	S. 52° E.
Westernmost ditto . . .	=	S. 64 E.
Temple farther south, of Thothmes III.,	=	S. 51 E.
Ditto ditto, Rhamses II.	=	S. 44 E.
Ditto ditto, Rhamses III.	=	S. 43 E.

ON EAST SIDE OF NILE.

Karnak, great temple, . . .	=	N. 64° W.
" grand avenue, . . .	=	S. 19 W.
" smaller avenue and temple,	=	S. 26 W.
Luxor, north part of temple of Rhamses II.	=	N. 40 E.
" south part, or of Amenophis III.	=	N. 30 E.

Besides these temples, all of them large, there are many smaller ones more variously placed still ; while the tombs which honeycomb the hills over miles and miles of rock and ravine, are truly almost scattered at random.

A lamentable change is this, from the regularity and order observable on the ancient hill of Jeezeh, where the Great Pyramid faces the cardinal points so truly, that it is difficult to discern any error ; where the second, third, fourth, and down to the eighth and ninth Pyramids, closely follow the same emplacement, and are themselves succeeded in a similar manner by every tomb, large or small, on the slopes of the hill around them; and where even every narrow, though deep, sepulchral well, is square in plan, and its sides are duly oriented.[1]

Order, method, and regularity, with a close attention to the leading facts of astronomy, conjoined with an appreciation of geometrical excellence and mechanical exactitude,—rule over the first expressions of Egypt amongst its Pyramids near the Delta. But garish display, luxurious living, contempt of science, as well mathematical[2]

[1] This refers in strictness only to such of the buildings on the Jeezeh hill, as can be safely and securely identified with the earlier Memphian, or ante-Theban, dynasties. Buildings of other and far different dates are unfortunately amongst them, and require to be sedulously guarded against in any critical inquiry. These subsequent architectural growths connect themselves chiefly with the revival of Memphis, after the glories of Thebes had passed away, in the 26th dynasty; when the memory of King Suphis of the 4th dynasty was converted into a god; when similar honours were conferred on the Great Sphinx,—a work, as Mr. Osburn shows, commenced in the 14th dynasty; and when, or by which period, the whole figment of Osiris and Typhon had been fully elaborated, with all its endless processions of animal-headed gods and mythical demons, quite unknown to the earlier empire of the Pyramid-building kings.

[2] M. Renan, in *Revue des Deux Mondes*, April 1865, remarks on the inferior stones, badly chosen, and worse put together, as if by slave labour and under compulsion of the stick, in the over-

as astronomical, and idolatry rampant,—mark the second stage of Egypt's national existence on the fertile plains of Thebes.

adorned temples of Thebes, while he can never praise sufficiently the conscientious labour and mechanical excellence of the Great Pyramid's plain construction.

ARCHITECTURAL DATA—*continued.*

CHAPTER 6.

VALLEY OF THE NILE—*concluded.*

Pyramids of Ethiopia.

HIGHER still up the Nile we next ascend,—through all Egypt and into Ethiopia, or Upper Nubia beyond;—and then, in the latitude of 17° N., or near the cities or stations of Meroe, Barkal, and Nourri, more groups of Pyramids are met with.

But we have now descended the stream of time to the days of Tirhakah, King of Ethiopia, and conqueror of Egypt under the 25th dynasty; or to about the date 700 B.C.

Was it, that that energetic king had seen the Pyramids of Jeezeh in the north, and wished to repeat, without understanding, them, in his southern home? Or did he propose to *improve* on what appeared to the rich profusion of his Cushite taste, the pale meanness of the undecorated memorials of primeval time? Or again, did he desire to combine into one, the augustness of the Pyramids with the splendours of Thebes?

We know not the idea which actuated the royal builder, unless it be expressed in, and by, his works themselves; and there, we find Pyramids numerous and square-based, but small (*i.e.*, seldom more than 1000 inches long in the side of the base); and very badly built,—consisting merely of an outside casing of showy masonry, filled inside with veritable loose rubbish. They, *i.e.*, these little Ethiopian Pyramids, are likewise thin, steep in their sloping triangular sides, and slender, though often slightly flat-topped intentionally; they stand, too, in groups together, like cypress-trees in a Turkish burial-ground; and are adorned with pedestals, roll mouldings along their corner edges or arris lines; and with occasional miniature belfry-like windows, or circular[1] and other ornamental markings near the summit; while below, they are still further decorated with a temple kind of hollow propylæon or porch, attached to, and projecting from, one side.

[1] This beginning of circles put in at random, for apparently mere idle thoughtless ornamentation, is noteworthy, and will be alluded to again.

These porches are indeed often so rich and architectural, as to be neither more nor less than small Theban temples; with pylons and pylon-towers, and two or three successive diminishing courts. But their ceilings are characteristic; being formed rain-proof and complete, sometimes in a round, and sometimes in a Gothic, *i.e.*, a pointed arch, within. Nicely, too, nay even gorgeously, painted, are the interiors of these Pyramid appendages; generally in a blue ground scattered over with golden stars on the said ceiling; and upon the walls, with the animal-headed gods and mystic abominations of later Egypt,—as well as with full-length portraits, again and again, of the reigning Ethiopian monarch and his large-bodied, thick-lipped, negro wives.[1]

Astronomical Emplacement.

And now, amongst these many architectural indulgences, how about the asserted rigid law of astronomical orientation?

[1] Fergusson's woodcuts of these Ethiopian Pyramids, taken from Hoskin's work, hardly make enough of the porches. These are better, as well as far more abundantly given, in Dr. Lepsius's folio *Denkmaeler;* both as to general views of the buildings in the earlier volumes, and large-sized copies of the almost innumerable paintings on the internal walls, in vol. x.

Our excellent general authority, Mr. Fergusson, says,—that it is carried out among those Ethiopian Pyramids, with even more than Theban irregularity.

That opinion being founded apparently altogether upon Mr. G. A. Hoskin's travels in Ethiopia,[1] —I have checked it by referring to Dr. Lepsius's large work, descriptive of his subsequent visitation of the same places: and on correcting his magnetic bearings given in all the plates, for the amount of 9° variation which is entered in some, —the following results appear.

At Meroe, where the porch looks out, or enters in, at the centre of the side of the Pyramid, and at right angles to the base side of the same,—the direction of such look-out is—

For one, two, or more Pyramids, .	=	S. 44° E.
Another, ...	=	S. 70 E.
"	=	S. 73 E.
"	=	S. 74 E.
"	=	S. 75 E.
"	=	S. 80 E.
"	=	S. 85 E.
"	=	S. 89 E.
"	=	N. 82 E.
"	=	N. 76 E.

Similarly at Nourri,—and where, in one of

[1] A very good book, for an original explorer; quarto, 1835.

Mr. Hoskin's plates, no less than thirty-six Pyramids are represented,—they are all square-based, but only a portion are supplied with porches; and the look-out of these bears,—

In some,	. . . .	= S. 41° E.
Others,	. . . . . .	= S. 44 E.
And others still,	. . . . .	= S. 51 E.;

while at Barkal, where all the Pyramids are square based, and all are furnished with porches, the look-out of these is found to be,—

For some,	. . . .	= S. 31° E.
Others,	. . . .	= S. 39 E.
"	. . . .	= S. 51 E.
"	. . . .	= S. 57 E.
And a large temple close by,	.	= S. 53 E.

Hence, if there be any law at all, it is rather for pointing to as far as possible from, than close to, the cardinal points. So that, after a course of 1800 years, we find that the architecture of the valley of the Nile, though beginning in the north and ending in the south with Pyramids,—had entirely departed from the primitive characteristic of astronomical emplacement. At the same time, the structures themselves were so altered in figure, —that in place of resembling the combined simplicity of a mathematical, and organic mystery of a natural, crystalline figure,—the Ethiopian Pyramids were such, that,—only increase their already commenced deviations a little more, or *e.g.*, pull up the Pyramid portion a little higher, ornament its sides more, and slightly enlarge the hollow porch, —and you have a northern Christian church with its spire: or, raise the pylons in an equal degree and you have the two western towers of a Gothic cathedral.[1]

This idea may be a little overstrained; but, looking to the high importance which has been attached in some quarters, to Sir Gardner Wilkinson's demonstrations of the much *later* pointed arches of the Rhoda Nilometer near Cairo,—being yet several centuries older than the earliest known Gothic example of similar *pointed* arches in the north;—there may be importance in directing attention to the same forms occurring in Ethiopia, both in greater profusion, and full 1000 years before even the Rhoda example. The architectural problem too seems easier, to account for the first invention of the figure of our churches

[1] Mr. Fergusson had already remarked in his *History of Architecture*, p. 104, of the Rhamession at Thebes, built by Rhameses the Great—'Its façade formed by two great pylons or pyramidal masses 'of masonry, which, like the two western towers of a Gothic 'cathedral, are the appropriate and most imposing part of the 'structure.'

in that manner, (though it does imply an adaptation of Pagan forms to Christian purposes,) than to obtain it out of the doctrines, injunctions, or descriptions of early Christian worship contained in the New Testament.[1]

Last Days of Nile Architecture.

Beyond those just mentioned, only a few isolated antique structures in African deserts, are occasionally spoken of by some wandering travellers; —but even the best of such remains are taken for ever out of the course of our research, by the following condemnatory sentence from Fergusson's discriminating history:[2]—

'The ruins of Wady-El-Ooatib, a little further up the Nile 'than Meroe, should perhaps be also mentioned here, if only 'from the importance given to them by Huren, who thought 'he had discovered in them the ruins of the temple of Jupiter

---

[1] The principles of much Puginian colouring may also be seen in these Ethiopian pyramid-porches of 2500 years ago; especially in the favourite background of deep and pure blue,—when on the ceiling, heightening the effect of the golden stars; and on the walls, throwing forward the figures, human or otherwise, of the artistic design, tricked out always cunningly, more or less in warmest tints of yellow, orange, and brown.

[2] Excepting, what may be to be excepted, about the supposed Dr. Leider's Pyramid, several miles west of the Great Pyramid of Jeezeh. See *Life and Work at the Great Pyramid*, vol. i. p. 455, and vol. ii. p. 184.

'Ammon. They are, however, all in the debased style of the 'worst age of Ptolemaic or Roman art in that country. They 'are wholly devoid of hieroglyphics, or any indication of 'sanctity and importance, and there can be little doubt that 'they are the remains of a caravanserai on the great commer-'cial route between Egypt and Axoum, along which the 'greater part of the trade of the East arrived at Alexandria 'in the days of its magnificence.'

Hence we have now exhausted, for our purposes, all the architecture of the valley of the Nile; and far away indeed from these latter degraded examples of it, must we look to other lands,—if we would hope to alight on more of the really *early* traces of mankind.

ARCHITECTURAL DATA—*continued.*

CHAPTER 7.

VALLEY OF THE EUPHRATES.

Babel Remains?

VERY generally throughout the world runs a belief, that the valley of the Euphrates and the plains of Shinar, formed the first *habitat* of intellectual man; and an eloquent author whom we have already quoted (page 8), even goes so far as to assert, that the idea of Pyramidal structures, and the practice of orienting them astronomically, arose there.

If belief alone could be admitted in this inquiry, the Tower of Babel, as described in the Scriptures, might be brought up in part testimony; and would, if so discovered, be found with little doubt to be older than the Great Pyramid of Lower Egypt, the oldest example of all the architecture which has thus far crossed our path.

On the *contemporary* monumental examination principle, however, whereon we are now embarked, no mere beliefs can have any place. We must appeal rather to the positive excavations or ground surveys of Loftus, Layard, Taylor, Sir Henry Rawlinson, and other modern explorers; and are informed thereby, that though 'Babylon' be the reputed site, very nearly, of the Tower of Babel, and though many ruins have been discovered there,—they are all much more recent, and far less notable, than anything which could be worthily identified with that primeval tower of belief.[1]

Most unlike the Great Pyramid of Jeezeh, which, standing on the hard, solid, living rock, is still on its own sure foundations, with nothing perishably human below, and nothing elevated above it,—the Babylonian buildings are immersed, rather than founded, in all the insecurity of illimitable depth of alluvial mud; such as the Euphrates brought down ages ago, and such as it

[1] 'It is by no means impossible that the rich alluvial plain of 'Shinar may have been inhabited by man as early as the valley of 'the Nile; but if this was so, it is certain that the early dwellers 'in the land have left no trace of their sojourn which has as yet 'rewarded the search of modern investigators.'—Fergusson's *History of Architecture*, p. 130.

is bringing down still, with the effect of continually and even rapidly extending its banks into the head of the Persian Gulf. Hence all those Lower Mesopotamian buildings, seem more or less endued with a tendency to sink beneath the surface of the ground; and one looks in vain for the mighty walls of even mediæval Babylon; *i.e.*, the Babylon of Nebuchadnezzar, the city whose mural crown was one of the wonders of the earth; with walls two hundred cubits high (according to Herodotus), of a width at the top for seven chariots to drive abreast, and furnished with a hundred lofty gates of costly bronze.[1]

Much digging amongst the ruins has taken place of late years, and has discovered, *first*, many traces of historical Babylon; and, *second*, beneath them, has come upon the remains of a very much earlier Chaldæan monarchy. Beneath these still, could the excavations be pushed so far, we might perhaps expect to find the foundations both of the Tower of Babel, and that city of the same name which we are told was

[1] See Rawlinson's *Ancient Monarchies*, vol. iii. pp. 348, 349, for many varying accounts of the immense heights and breadths of these walls; and at page 338 for the testimony that 'no vestige 'of the walls of Babylon has yet been discovered.'

the beginning of the kingdom of Nimrod, did it ever really stand on the surface of the ground in that precise locality. But as no such foundations have yet been found or certainly identified,—we must for a time discharge even the memory of these ancient names from our collection of data; while we proceed to the consideration of whatever remains actually *have* rewarded the zeal of excavators in various sections of the Interamnian country, and are known now to archæologists as 'the Pyramidal temples and burial cities of 'Proto-Chaldæa.'

ARCHITECTURAL DATA—*continued.*

CHAPTER 8.

VALLEY OF THE EUPHRATES—*continued.*

Proto-Chaldæa.

THE chief period of activity and importance in the Proto-Chaldæan kingdom, seems to be reckoned by the Rev. George Rawlinson, in his *Five Great Monarchies of the Ancient Eastern World*, as ranging between chronological limits of 2200 B.C. and 1500 B.C.

But then as his date for the Great Pyramid of Jeezeh, though far from so extravagant as those of many other literary men, is yet some 300 years greater than that derived from the only irreproachable astronomical data,—we shall probably be justified in slightly diminishing his earlier numbers; and calling *his* 2200 B.C., 2000 B.C. astronomical. In this manner his dates will fit symmetrically into those already ascertained for

Egyptian history; and will not be materially different from those approved of by Sir Henry Rawlinson, Mr. Fergusson, Mr. Loftus, and other names of authority in Mesopotamian history.

The first supposed or traditional monarch of this old Chaldæan empire, according to the chief author, just quoted, was Nimrod; but the earliest *proved* one Urukh, whose name has been found inscribed on many of the cuneiform inscriptions; which were prepared and inserted in his buildings by himself, contemporaneously with their erection; and consequently with every admitted and recommendable attribute of efficient testimony to their architect and paternity. Such testified remains have been found moreover at several different localities; though only at two of them have any exact traces for our present inquiry been discovered; viz., at Mugheir in lat. 30° 57′ N., with long. 46° 12′ E.; and at Warka, supposed to be the ancient Erech, in lat. 31° 20′ N., and long 45° 40′ E.

The chief practical explorers in this important field have been Sir Henry Rawlinson,[1] Mr. Loftus,[2]

[1] *London Asiatic Society's Journal.*

[2] And his own octavo volume, entitled *Chaldæa and Susiana.*

and Mr. Taylor;[1] and the principal discussers of observations — Rev. Dr. Hincks, Rev. George Rawlinson, Mr. Fox Talbot, and Mr. Fergusson. All these gentlemen, too, seem to be pretty well agreed, that the buildings just cited and evidently chief ones of old Chaldæa, were of a temple, or religious-worship, character; having, towards that purpose, a small rectangular hollow tower or chamber, based on a larger mass constructed like a terrace; and this on a yet larger one, or a sort of foundational mound of earth, strengthened outside by more or less amount of builded casing,—an idea, in architecture, which is said to have originated in the days of King Urukh, and perhaps to have been invented by himself.

The best materials employed, being seldom more than bricks; and these usually crude or unbaked,[2]—the buildings they composed, have in the long subsequent ages mostly crumbled down into rounded mounds of mere earth, which cover all detail with their own featureless decay. No-

[1] *London Asiatic Society's Journal*, vol. xv.

[2] Some of the better preserved masses of brick-work, are said to owe that superiority to the introduction of a layer of reeds and rushes between every few courses of the crude bricks; preventing thereby rain from cutting and washing out deep ravines with the same facility that it does in pure clay.

thing accurate, therefore, was known in modern times, touching the said fiducial details, until the explorers above named commenced extensive excavations into the said mounds; and, after cutting deeply through the superincumbent mass of clay and soil, reached some portion of the actually standing lowest walls.

Such structures, then proved to have been fairly rectangular as to their angles in ground-plan, but decidedly oblong as to their sides. The lowest terrace (for the formation was always in such terraces, decreasing in size as they ascended) invariably exhibited enormous external walls, which were either deeply pannelled, or provided with enormous buttresses, giving both strength, stiffness, and variety of light and shade under a bright sun. Upon this was placed a second terrace, smaller, of more refined architectural features, but oblong also in its rectangular ground-plan, and eccentrically situated as to the lower terrace. On this second terrace was placed sometimes a third one, of similar order; and on the top of that again, a small hollow apartment or shrine, exceedingly ornamented, often with precious metal, but also founded in eccentric position.

These several eccentricities were all so arranged, as to push every successively higher terrace or tower more and more towards one, than the other, end of the general mass. One end of the same being thereby made to rise from the plain (though, still by the terrace steps), more, and the other less, steeply,—the greater roominess of the latter end was utilized by having large flights of steps, variously and showily arranged, so that whole crowds of worshippers might, if necessary, ascend and descend with all comfort and decorum.

The sides of these terraces were generally arranged at a large angle of slope, or convergence inward, as they ascended; but that was all they had of the true Pyramid shape; and there were no nearer approaches to that most peculiar, and easily definable, form throughout all Proto-Chaldæa.

Oblong rectangles then in plan, built in successive eccentric stages, with unequal sides of trapezoid shape, protruding with buttresses, deeply cut into with staircases, and surmounted by a flat-topped hollow chamber,—the Proto-Chaldæan temples evidently were no rivals to the mathematical perfection of realization shown by the

Pyramids of Lower Egypt. 'Pyramidal temples,' we may go on calling the former, because the fashion has been set by men of mark; but only distantly pyramidal are they; while the name of 'Pyramid' would be a flagitious misuse of the term.

Picturesque, rather than scientific, forms, must these structures of the Lower Euphrates have been. Even in their ruins, travellers speak of them as affording the only varied point for the weary eye to rest on, in the trackless, featureless, round of the sharply-defined, level horizon, terminating all views in the flat and ocean-like alluvial plains or marshes of the Interamnian region. To direct their camels' march on these distant artificial mountains, is the delight of Arab drivers; and it is the pride of European artists, on a nearer approach, to adopt every allowable device, such as the shadow of a passing cloud, or the rays of the sun bursting out behind some partially eclipsing screen,—to throw a glory which is not their own, upon these much-meaning mounds of old, but where now only a few gaunt cattle crop a little grass after the winter rains are over, and before the fiery heats of spring commence.

Still more of this picturesque effect must these structures have had in the days of their youth; for, taller and larger then, they supplied more completely the natural want of mountains in a flat land. With their serrated, terraced edges, too, and greater steepness on one side than the other, these Chaldæan temple masses possessed all that variety of outline which connoisseurs in hill scenery so much insist upon; while the metallic decorations of the upper tower or chamber, must have reflected the rays of the rising or setting sun as brilliantly as the snowy summits of the mighty Ararat,—the 'terrible crystal' of Ezekiel's story.

So much, then, for an architecture which never had any claims to geometric praise. But can better things be said for its astronomy?

Astronomical Emplacement.

Alas! a yet more remarkable difference awaits us here than any hitherto touched upon; for, Proto-Chaldæan templar edifices are always placed, so say all our authorities, not with their *sides*, but their *corners*, towards the four cardinal points!

This statement is repeated so often and so positively by both Loftus and Rawlinson, that they would seem to have overlooked the impossibility of *all* the corners of a long rectangular building being directed truly to the four cardinal points, and to there being a difference of many degrees to be disposed of somewhere, as error on two corners, if other two had been set true, as above indicated, from one central point, on a ground-plan of the very long rectangular proportions which they give.

The better practical method of proceeding would probably have been, for our explorers to have recorded the direction of the *sides* of a structure (eliminating the buttresses); and then, in stating that *they* (the sides) point half-way between all the four cardinal directions, there would have cropped out more prominently the astronomical antithesis of these Chaldæan buildings to all Egyptian Pyramids; viz., that instead of being oriented, as those are, *to* or *upon* the cardinal points, they are as *far from* them, angularly, as they possibly can be.

That at least would be the case, did the walls run accurately at 45° from those directions; which is what has been implied for them, and even claimed positively for the governing intention of the builders. But within what limits of accuracy are the temple remains actually found to do so?

Mr. Loftus's plate of Mugheir, as closely as I can test it, gives S. 37° E., or N. 37 W., for the direction of the two larger parallel walls; and another for Warka gives N. 39° E., or S. 39° W.; the entry to the former being certainly represented on the south-eastern side, and of the latter apparently on the north-eastern. In either case very different from the Egyptian Great Pyramid's entrance, from due north; and its error less than 5′, in place of these 8° and 6° fallings-short on what was desired and intended, whether religiously, scientifically, or symbolically, in these earliest known beginnings of Lower Mesopotamia.

But for whatever purpose the temples were designed, there can be no doubt what the neighbouring cities were used for; viz., as places holy, or gifted with some peculiar virtue, for burial, and resorted to on that account by large populations from even distant countries and through many ages. Graves, graves, graves, therefore, fill all the environs of these primeval cities. Not costly tombs like the Egyptian grand and far-between sepulchres, but every kind of compendious coffin made in baked and unbaked clay; perfect models of gigantic earthenware, stuffing the ground as full as it can hold on every side.

The same idea still lives in the same region, as shown by the pertinacity with which all the rich throughout Persia send their dead in ceaseless streams throughout every year, to be buried within the precincts of the sacred city of Kerbela.[1] A Mohammedan martyrdom of Hassein, son of Ali, sanctifies the practice immediately in their eyes; but the custom is evidently inherent in the region, and dates from a period up to which no historian has ever penetrated.

[1] Lat. 32° 40′ N.; long. 43° 40′ E.

ARCHITECTURAL DATA—*continued.*

CHAPTER 9.

VALLEY OF THE EUPHRATES—*continued.*

Assyria.

AFTER the decay of that most ancient, or Proto-Chaldæan kingdom, the seat of Interamnian empire appears to have passed north-westward up the Tigris; or to that region where Layard, in the mounds of Nimroud,[1] uncovered what he thought at first to be Nineveh,—but which is now recognised as ancient Calah; and where M. Botta at the same time uncovered in the mounds of modern Khorsabad[2] what he also thought must have been Nineveh, and published as such,—but which is now generally recognised as Bit-Sargina; and where Layard again in subsequent years excavated at Koyunjik,[3] what appears to have

[1] Lat. 36° 9′ N. and long. 43° 26′ E. approximately.
[2] Lat. 36° 35′ N. and long. 43° 18′ E. "
[3] Lat. 36° 22′ N. and long. 43° 16′ E. "

been the true Nineveh: all of them, however, being cities, and very splendid ones too, of that one Assyrian empire which flourished under well-known kings from about 1600 B.C. to 625 B.C.

An allied race to the old Chaldæans, would the Assyrians appear to have been from their architectural methods,—though notably differing in their objects; which indicate a ruling tendency to make all their greatest buildings palaces for their kings, rather than temples to their gods.[1] Comparatively little did they care for the glorification of their dead, whose tombs nowhere appear; and indicate thereby more of the Semitic, than the Turanian, element of race,—according to the Fergussonian principle. In manner of construction too, they aimed at more use of stone; large supplies of which, in shape of alabaster slabs, were comparatively close to them in the Assyrian hills.

In the matter of orientation, the greater part of the buildings at Nimroud, particularly those of

[1] At p. 169, Mr. Fergusson speaks of the *pyramid* of 'Nimroud,' but a little further on he allows that it rises in successive terraces like the lower Chaldæan temples; though he does not see evidence either of temple, or tomb character in this one pyramidal building at Nimroud.

the north-west palace, are represented by Layard as having their walls directed to the cardinal points; although those of his south-east palace are placed diagonally. At Khorsabad, again, in M. Botta's fine plates, the walls of *all* the palace chambers invariably run diagonal fashion; or, as measured on one of Mr. Rawlinson's plates, N. 41° E., with S. 41° W., and N. 49° W., with S. 49° E.

While finally at Koyunjik, or Nineveh itself, the palace walls in Mr. Layard's plates all cross the cardinal directions; and in one of the most measurable instances, at angles of N. 67° E., with S. 67° W., and N. 23° W., with S. 23° E., while another indicates N. 60° E. with its consequences.[1]

No great accuracy therefore in keeping to anything astronomical is manifested by the buildings of the great empire which terminated with the fate, classically, of Sardanapalus, or, according to the cuneiform inscriptions, with Asshur-emid-ilin; but they, the builders, decidedly were against the asserted theory of accurate orientation *with* the cardinal points; and indicate very plainly on the whole, a desire to fall in with the opposite idea,

[1] See Layard's *Nineveh and Babylon*, 8vo, 1853; also his *Monuments of Nineveh*, folio, 1849: and a second series in 1853.

or that already produced on the lower Euphrates; viz., of making their walls point azimuthally as far as possible from the said earth-axial directions.

Magnificent restorations of these Assyrian palaces have been published by the accomplished Mr. Fergusson, enabling us to realize much of their highest character and their æsthetic effect on the mind, by virtue of form, size, and chromatic, joined with sculpturesque, decoration.

'Judged,' he says, 'by the same rules of criticism which 'we apply to classic or mediæval art, the architecture of the 'Assyrians must, it is feared, rank very low. But for gor-'geous barbaric splendour of effect, it seems difficult to imagine 'anything that could well have been grander or more impos-'ing than the palaces of Nineveh must have been, when entire 'and filled with the state and magnificence of the monarch of 'Assyria.'

To acquire a notion, too, of the condition of mind of these worthies, we may refer to the very words which one of these kings himself recorded, about 950 B.C., as translated by Mr. Fox Talbot, (the inventor of British photography,) thus:—

'Within the circuit of this fortress I constructed a fine 'house of cedar; a house of cypress wood; a house of *taprani* 'wood; a house of *ker* wood; a house of *meshkani* wood; a 'house of terebinth wood and of *tarpikhi* wood, for the resi-

'dence of my majesty, and for a remembrance of my reign 'for evermore.'

'I made sculptures of the animals of the lands and seas, 'carved in *pari* stone and *paruti* stone, and I set them up at 'the doors of my palace.'

'I made it grand; I made it splendid, and with images of 'bright copper I adorned it.'[1]

Thus, indeed, things went on at Nineveh. There was war and the slaughtering of captives; building of palaces, and taskmasters to smite the builders at their tasks; the chase for the king's diversion; the many ladies of his hareem for his entertainment, and a continual wine cup-bearer at hand, the most honoured official of all his officers, to keep him constantly served up to the highest stage of physical enjoyment. How, then, could *science* enter there! Not, indeed, that science is a forbidden inquiry or pursuit to kings; for a century before Asshur-akhbal flourished, a wiser monarch than he had written in another land,—

'It is the glory of God to conceal a thing; but the honour 'of kings is to search out a matter.'

The kings of Assyria, however, in their infinite pride of never-ending luxury, cared not to search

[1] Assyrian inscription of Asshur-akhbal, a monarch of the tenth century B.C.—translated by H. Fox Talbot, Esq.; p. 202, vol. vi. part 1, of *Proceedings of the Society of Antiquaries of Scotland.*

out any of those difficult or wondrous, or soul-instructive secrets of nature, which now form the bulk of modern science; and which were concealed in the beginning by the Almighty, only in such manner as to place the rudiments of discovery within the range of man, if he will but exert himself honourably and mentally.

Science, therefore, lost nothing when Nineveh fell; though mere royalty, based on slave power and luxurious living, received a heavy blow and dire discouragement, in the funeral pile which was lighted on a fatal day in the treasure-house of the grandest of Assyrian palaces. Most complete, indeed, was the destruction then wrought: so that the earth itself refused to tell, through 2500 years, the place where that monarchy, so useless to the progress of mankind in general, had flourished so proudly and so long; and whatever advances men have made since then, whether in science or art, in war or peace, in religion or morals, they owe little indeed to the subjects and descendants of Asshur the Great.

But something better, it is often affirmed, can be said of the neighbouring city of Babylon. By all means, then, let us see.

ARCHITECTURAL DATA—*continued.*

CHAPTER 10.

VALLEY OF THE EUPHRATES—*continued.*

Babylon.

BABYLON, which had long been in existence,[1] and slowly growing through a chequered career of alternate independence and subjugation, revolt and certain defeat,[2] seems to have sprung up into immediate and magnificent sovereignty about 625 B.C.; that being the date of the destruction of Nineveh, Babylon's formidable rival of old, and even tyrannic oppressor, in the north-west.

[1] 'The original Chaldæan monarchy perished through an Arab 'invasion about 1500 B.C., and the invaders held possession of the 'country till 1273 B.C. Their rule was then superseded by that of 'the Assyrians, who became masters of Babylonia under the first 'Tiglathi Nin, and governed it for a short time from their own 'capital.'—Rawlinson's *Ancient Monarchies*, vol. iii. p. 469.

[2] 'The middle period of Babylonian history, the time of obscurity 'and comparative insignificance, when the country was, as a general 'rule, subject to Assyria.'—*Ibid.*

Babylon therefore then, and from thence, formed for itself the second Chaldæan kingdom, and nearly on the site, though rather to the west and north of the more ancient one; or with a central latitude of 32° 32′ N., and longitude 44° 28′ E.

This being not far from the position assigned by general tradition to the scriptural Tower of Babel, the more modern Babylon,—besides its own latter-day glories of palatial, templar, and mural architecture,—inherited the fame, not only of a previous race of truly ancient Chaldæans, but of the most primeval men, in post-diluvial times, of all religious story as well.

The Arabs even point to an isolated, extensive, and now rather flat-topped hill of rubbish, lying north of the veritable palace mounds of Babylon, and, calling it Babil, declare that it is the remnant of the very Babel Tower itself. Modern excavation researches, however, have shown it to be more probably that temple of Belus which was completed by Nebuchadnezzar about 605 B.C., or four centuries later than Solomon's Temple at Jerusalem, and one century after the Ethiopian Tirhakah's attempted resuscitation of burial Pyramids on the banks of the Upper Nile.

Of its latter history, M. Oppert writes:—

'The successors of Nabuchodonosor occupied themselves
'equally with the embellishment of the Pyramid and Tower.
'Xerxes, on returning from Greece, decorated the Pyra-
'mid, which, according to the historians, contained the tomb
'of the god. Ctesias and Ælian report that the king of
'Persia, greedy of treasures, found there a sarcophagus
'half full of oil. Alexander employed, during two months,
'10,000 soldiers to open up the ruins of the Pyramid. A
'little time after, the great Macedonian died, and with him
'was interred the idea of the reconstruction of this ancient
'monument.'

Besides this mention of the tower as originally a Pyramid, M. Oppert, in one of his restored maps of Babylon, copied by Mr. Rawlinson, represents the building in plan, as a perfect mathematical, quadrangular-based Pyramid, having four triangular sides, whose vertices meet in a point above. But this shape is most absolutely denied by the Bampton Lecturer, who insists that recent examinations have proved that the structure was similar to most or all other Mesopotamian temples, in having consisted of a series of terraces eccentrically placed, and diminishing in breadth as they ascended; while Mr. Fergusson considers that its base may have been a parallelogram of about 600 feet square; for although some measures have given 657 feet for the south, and 408 for the west side, he describes it as being too ruinous for anything more than the merest approximation to its original size or shape.

Hence, to form any certain idea of the templar architectural style of neo-Chaldæa, we must proceed to other examples; and the best of these, known chiefly by its Arabic name, 'Birs-i-Nimroud,'[1] lies some twelve miles south of the city of Babylon itself, and on the opposite side of the Euphrates.

The Birs-i-Nimroud Tower.

In this most notable mass, Sir Henry Rawlinson excavated, and discovered not only the commemorative inscribed cylinders of King Nebuchadnezzar himself, but a sufficient length of several of the walls to determine that most important, yet often utterly-disputed, point of the astronomical emplacement.

This building, too, is all the more appropriate for our present inquiry into a reported astronomical *condition* of Pyramidal and general

[1] To be carefully distinguished from the Assyrian Nimroud, the scene and the making of Layard's early fame.

arrangement; for, while it is fully as close an approach to a pyramidal building as anything ever known to have been erected in Mesopotamia, it was also intended by its founders and finishers to be an astronomical building; viz., the temple of Nebo at Borsippa, dedicated to the seven planets or heavenly spheres of Babylonian astronomy.

In fact, if we may trust the findings of recent literary research, we have now to deal with, both in this pyramidal temple, its architect, and its builders,—yea, indeed, not only to deal with, but perhaps be guided by—the most serious, and certainly most promising, case, which has yet come before modern astronomers in the recognised history of the learning of the world. For, of the Babylonians of those times, writes the Rev. Mr. Rawlinson,[1]—

'Inheriting a legacy of scientific knowledge, astronomical 'and mathematical, from the proto-Chaldæans, they (the 'Babylonians) seem not only to have maintained, but con-'siderably advanced, those sciences by their own efforts. 'They were good observers of astronomical phenomena, care-'ful recorders of such observations, and mathematicians of no 'small repute.'

[1] P. 328, vol. iii. of *Ancient Monarchies.*

Nay, he even transcends everything that has ever yet been established of ancient science, by stating[1] that some of the Babylonian cuneiform inscribed cylinders, now in the British Museum, seem to indicate, so far as they have yet been interpreted, that the four satellites of Jupiter and seven of Saturn had been observed in Babylon; a proof, he considers, that the small and simple lens of rock-crystal discovered in an earlier Assyrian palace, had developed in the hands of the sages of the southern city, into a telescope; and a surpassingly good telescope too, even in our days of achromatic glasses, every modern practical astronomer will allow, if seven satellites of Saturn had really and certainly been observed through it.

All these things, however, depend on literary interpretations, wherein I am no authority; or on literary histories, which, being by no means *contemporary* with the events they assume to decide, can have no effect in our present work. Let us therefore turn instead to the positive accounts from modern examinations, and especially respecting that feature at once so practical and yet so refined; viz., the astronomical orien-

[1] P. 424, vol. iii. of *Ancient Monarchies.*

tation of the existing walls of this temple of the heavenly spheres.

Astronomical Emplacement.

As with many other remnants of Eastern antiquity, a *cardinal* direction[1] had long been assigned by our earlier travellers to the walls of the Borsippa building; and assigned because there was apparently a foregone conclusion that it must be so, and could not be otherwise. In most of the Babylonian cases, extensive ruin has prevented either proof or disproof; but at Borsippa, Sir Henry Rawlinson, after having made far more extensive explorations and careful measures on the spot, than any of his predecessors, repudiates the cardinal idea in these most positive terms:—

'I must here observe that Rich and Porter have both been 'guilty of a most singular error in describing the sides of the

---

[1] Such a direction has always been assigned by the Rev. G. Rawlinson to the walls of 'Babil;' and it *may* be so; but we must certainly object to his note in p. 367, vol. iii. of his *Ancient Monarchies*, where, while he allows that such a direction would be anomalous amongst the usually *diagonal* astronomical emplacements of Babylon, he disposes of the anomaly by saying, that for the astronomical purposes which the towers subserved, it was 'indifferent 'which arrangement was adopted.' We can only hope that at the Oxford Observatory, it is *not* a matter of indifference whether the meridian circle is parallel to the meridian, or 45° therefrom.

'Birs as facing the four cardinal points. In reality, it is the 'four corners which, with a slight error, face these points; 'and the titles of Kerr Porter's plates must be thus altered 'throughout the series; his western face being S.W.; southern 'face, S.E.; eastern face, N.E.; and northern face, N.W. The 'north-east face is the front of the temple (and enters with 'stairs), the south-west the back, and the other two the 'sides.'[1]

The Birs-i-Nimroud, from plates, both by Rawlinson and Fergusson, is evidently much more nearly square than the proto-Chaldæan temple of Mugheir; but yet its successive terraces are eccentrically situated upon each other,—though the sides are apparently parallel or intended to be so,—and it is not, therefore, the most correct method of describing its astronomical emplacement, to speak of the orientation of the *corners*; moreover, it is not so easy to observe accurately on the corners, as on the sides, of an ancient structure, as any one would soon find out on trying. Sir Henry Rawlinson too does not seem to have proceeded in that manner at Borsippa; but to have actually observed the trending of the sides, and then concluded from them for the corners; being perhaps involuntarily moved to

[1] *Journal of the Asiatic Society*, vol. xviii. p. 5.

give them so much prominence, because the glorification of the *corners* of a building was a peculiarly Babylonish method: and in two of the corners of this very building he had just discovered, to his own great fame among the Arabs, the commemorating cylinders of King Nebuchadnezzar.[1]

That king, however, had in this case only completed a structure previously begun by another monarch; or, as Sir Henry puts it, the foundation platform 'was left untouched by Nebuchadnezzar' when he rebuilt the upper stage.[2] And then follow his (Sir Henry's) important remarks on the orientation, determined fortunately, as already stated, by observation of the trending of the lines of the walls. Observed indeed only with an azimuth compass, but this seems to have been efficiently used; corrected for its errors and variation; and mounted for the time upon sharp

[1] 'That must be a very powerful instrument for finding treasure,' said one of the Arabs, pointing to the azimuth compass, wherewith Sir H. Rawlinson had taken a bearing from the corner of the wall, just *before* a further digging into it discovered the cylinder. And Sir H. R. was inundated immediately after that, with applications to assist both Pashas and Bedouins to find treasures of gold,—discoveries more after their hearts than any inscribed cylinder.

[2] *Journal of the Asiatic Society*, vol. xviii. p. 13.

edges of the wall, only recently excavated clear of the rubbish which had covered them for 2500 years.

Hence we may put more trust in the following, than in any general, statements about the orientation of any other Mesopotamian building whatever :—

'A curious illustration of this difference of age (at Birs-i-'Nimrud) is also found in the varying direction of the lines 'of brick-work, as occurring in the foundation and in the 'temple which it supported, the corners in the upper building 'nearly facing the four cardinal points, while the lines of the 'sun-dried bricks at the base are deflected 16° to the east. It 'is impossible, of course, that this great discrepancy between 'the two designs can have anything to do with astronomical 'variation; but for the small error from the true bearing, 'amounting to 4° or 5° which is apparent above, a natural 'explanation may very well be sought. We may assign the 'error, it is true, to imperfect instruments, but I should 'prefer explaining it by supposing the lines to have been 'laid on a day when the sun had 4° or 5° of eastern amplitude.'

And then follows in a note,—

'Mr. Fresnel gives the error from the cardinal points at 5° 'or 6°, and supposes this to be the magnetic variation of the 'spot. The true magnetic variation, however, at Babylon '(July 1853), determined by a series of azimuths, is 4°. The 'compass which I used had an error in itself of 1° the other 'way; and as my magnetic bearing was 52° 5′ for the line 'of the S.E. face, I thus gave the true error of the building 'at 4° 5′ E.

Babylonian Generalities.

Hence the actual contemporary facts of their astronomical buildings, do not speak much in favour of the practical skill of the Babylonian astronomers; nor indeed of their theoretical intentions either, as evidenced in the diagonal emplacement of all these walls, which, in a cardinal direction, might have aided many exact determinations of the places of stars and planets.

But the truth is, probably, that pure science was no more thought of, for its own sake, in Babylon, than was pure, self-denying, righteous religion. Even some of their best modern friends confess that

'the Babylonians were astrologers, no less than astronomers, 'wont to expound dreams, and to foretell events by means of 'the stars. The ancient writers, Biblical and other, state 'this fact in the strongest way; and the extant astronomical 'remains distinctly confirm it. The great majority of the 'tablets are of an astrological character, recording the supposed influence of the heavenly bodies, singly, and in conjunction, or in opposition, upon all sublunary affairs, from 'the fate of empires, to the washing of hands or paring of 'nails.'[1]

[1] Rawlinson's *Ancient Empire*, vol. iii. p. 425.

Their fine art, whether painting or sculpture, though evidently derived from the more or less grand conceptions of the Assyrian artists of the earlier time, degenerated amongst the Babylonians to the most puny, wretched, and contemptible representations of humanity,—whether on their gems, or clay cylinders, or the few colossal figures known (see especially their short-legged lion, with the thick tail; and the shapeless man under him), —that were ever perpetrated by a civilized people.

Painted by themselves, the rich Babylonians certainly look like what one might expect of an upper ten thousand, who fed their poor chiefly on 'pickled bats;' and who indulged themselves in magnificent banquets that 'generally ended in drunkenness.'

'Not that these were scenes of coarse indulgence,' says one of their apologists,—'for bands of performers entered with 'the wine, and entertained the guests with the soul of music; 'and rich odour of perfumes floated around; the eye was 'delighted with the display of gold and silver plate; while 'the splendid Babylonish vestments of the guests, the exquisite carpets and hangings, the numerous attendants, gave 'an air of grandeur to the scene, and seemed half to excuse 'the excess of which too many were guilty.'[1]

[1] Rawlinson's *Ancient Monarchies*, vol. iii. p. 329.

This implies wealth, and wealth there was in Babylon, for its pursuit was followed with all the keenness and avidity which has become the evil characteristic of the Jews of latter-day time; descended now from an ancestry partly mixed with Babylonians, and partly with Edomites, since the days of their greatest known kings.

But the Semitized Babylonians were still more unscrupulous, and absolutely stuck at nothing which should be thought good for the immediate profits of trade, and the attraction of moneyed strangers to the place. Whence arose the vilest feature of all in their degrading customs; and which rendered virtue an impossibility for every Babylonian woman, whether rich or poor; making the vice even a religious duty, and setting apart a special temple for the purpose. That temple, too, Herodotus contemporarily relates from his own observation, was always crowded with those coming or leaving,[1]—yet now its place knows it no more, and the whole city has become mere 'shapeless and ruinous heaps.'

This thoroughly corrupted city's statues, consisting often of plates of gold laid on a structure

[1] Rawlinson's *Ancient Monarchies*, vol. iii. p. 465.

of rotten clay, deceived and dazzled many in days of old. Even now, one author, though too, an English clergyman, has recently written, in termination to a general chapter of special pleading in favour of almost everything Babylonian,—that

'Babylon was the source to which the entire stream of Eastern 'civilisation may be traced.'

'To Babylon, far more than to Egypt, we owe the art and 'learning of the Greeks.'

'And it is scarcely too much to say that, but for Babylon, 'real civilisation might not yet have dawned upon the earth.'

But though these whitewashing opinions have recently emanated from Oxford and been printed in London,—and though certain practices connected with numbers and sexagesimal division, and certain prophetical almanacs still existing amongst us, may be traced for their origination up to Babylon,—yet everything Babylonish that can be tested by true science, is evil continually; a thorn in the side of honest philosophy, and an obstacle to healthy progress. Certainly thus do we find it with the astronomical emplacement of their buildings—and certainly thus may all men prove it to themselves for true civilisation, as dependent on a pure and true religion.

ARCHITECTURAL DATA—*continued.*

CHAPTER 11.

VALLEY OF THE EUPHRATES—*concluded.*

Media and Persia.

MEDIA occupied the hilly country north-west and north of the Euphrates and Tigris, but frequently descended with its immigrants or armies into the great valley of the plain of Shinar,—so as to acquire much historic connexion therewith. To Media also, some historians are inclined to give a special appearance on the scene of human actions from 2458 B.C. to 2234 B.C. (probably 2100 to 1934 B.C., true and astronomical); and then a long oblivion until it rose to notice once again in 835 B.C., just before its junction with the Persian rule.

But these earlier times seem to be merely mythic and traditional; for Ecbatane with its coloured walls, and all the other cities of the

'Medes' that have yet been discovered, are identified only with the later historic period; have furnished no data for their orientation; and, even if they had, can afford, from their comparatively recent dates, no *contemporary* proofs of the practice in vogue amongst the '*origines*' of mankind, in these parts.

Persia.

Similarly of Persia, occupying the country north and north-east of Shinar proper,—as it only flourished imperially and architecturally from Cyaxares in 538 B.C., to Darius Codomanus in 336 B.C.—there is little use in referring to any of its known building remains, for proofs of primeval orientation.

Towards the beginning of the present century the ruins of Persepolis, Shushan, and Passargadæ used to be regarded as some of the grandest and most mysterious remains of ancient times. But Mr. Fergusson's *Architectural Philosophy* has reduced them to their true proportions.

They are merely the forms, he says, of older Babylonian and Assyrian buildings, erected in a more durable material, or stone, in place of bricks, wood, metal and clay,—that use of stone, having been learned by the Persians, in their expedition to Egypt under Cambyses in 525 B.C.

But the Persians had then entirely ceased to be the energetic, virtuous, mountain-people they had appeared under their first great ruler Cyrus; and what they erected under the successors of Cambyses, with the assistance of degenerate Egyptian art, was little else than sumptuous piles of palace apartments; fit for their barbarian monarchs to revel in, and not unfit for Grecian Alexander to reduce to ashes.

The wealth and the military, as well as slave, power of Persian monarchs caused the buildings to rise from the ground. And thousands of prisoners of war, led along from their distant homes, stripped naked, with their hands tied behind them, and connected together in long lines by a rope attached to rings passed through their noses, or knotted on spiked collars fixed round their necks,[1]—were glad to exchange such corporeal pains as these, for the wasting labour of masonry and earth work, even under barbarian task-masters.

But such treatment could not procure genius;

[1] See Plate in Rawlinson's *Ancient Empires*, vol. iii. p. 436.

and the sculptured adornments of these ill-founded palaces, were only debased copies of older Assyrian designs; or mere mechanical repetitions of trifles wearying to behold. Some apologists have indeed appeared, for the often met group at Persepolis, of the lion and the bull. Such a lion as that, they say, so tremendous in power, was never seen in the best representations of Nineveh or Calah.

No, indeed! for who ever saw a lion half as large again and three times as heavy as a bull? But to represent *such* a lion, leaping on the poor bull, and tearing away visibly with all its teeth, and two great spreads of fore-feet claws all working at once into the tender flesh of the hind quarter of the bull,—who, poor animal, can only raise up its fore-feet in anguish and look round dismayed at its ferocious enemy;—to make such a representation of unnecessary cruelty as this, and to have so much pleasure in beholding it, as to have it repeated again and again, on every staircase, and almost every spare portion of the palace walls,—*that* surely argues a people culminating in themselves all the harder-hearted features of the several preceding nations of the Interamnian valley. For while they were, amongst themselves, 'given to pleasures,' 'exceeding in dyed attire 'upon their heads, painting their faces, wearing 'ear-rings, and clothing themselves in soft and 'rich material,'—yet were they to others, bitter and hasty, 'terrible and dreadful, and more 'fierce than evening wolves; smiting the people 'in wrath with a continual stroke, the hammer of 'the whole earth;' transplanting conquered races, mutilating prisoners, and massacreing non-combatants with all refinement of violence and cruelty, beyond even the ordinary Asiatic limits.[1] And finally, they were themselves at last extinguished by a perpetual judgment, which makes it needless for us to look among the subsequent inhabitants of the region, for any direct continuations of the alleged early astronomical science in the valley of the Euphrates.

[1] Rawlinson's *Ancient Monarchies*, vol. iii. p. 333.

ARCHITECTURAL DATA—*continued.*

CHAPTER 12.

CIRCULAR BUILDERS.

WE have been thus long and full hitherto, in tracing each of the two systems of architecture, the Egyptian and Euphratean,—from their first known examples, about 2000 B.C. or more, down to their loss or degradation a few centuries before the Christian era,—because they are, in the opinion of those best calculated to judge, the two which ascend up most nearly to the *origines* of intellectual man. Thus, Mr. Fergusson, after freely giving to the Pyramids of Egypt a decided pre-eminence in time, over whatever buildings have been discovered as yet in the valley of the Euphrates, says of the latter,—

'There is nothing, certainly, in India, that approaches these 'monuments in antiquity, nor in China or the rest of Asia; 'and in Europe, whatever may be maintained regarding prim- 'eval man, we can hardly expect to find any building of a date 'prior to the Trojan war (1190 B.C.) All our histories must 'therefore begin with Egypt and Assyria. Beyond these all 'is speculation, and new fields can hardly be hoped for.'

Yet, if we ourselves are not mistaken, there is a third variety of architecture, which should come, in almost equal chronological rank with Egypt and Assyria, and share their admitted claims to the honour of descending from very ancient times. This third architectural division, too, is abundantly distinguishable by specialities of type.

The two commencements of each of the River systems we have already noted, viz., the Great Pyramid of Egypt on one hand, and the pyramidal temple of proto-Chaldæa on the other, were sufficiently distinguished from each other in geometry; the former, being a pure mathematical figure of regular shape and easy definition; whereas the latter, was an unequal-sided structure, built in broken and eccentric terraces. They were also clearly separable by astronomy; the former being oriented cardinally, the latter, at the utmost opposite thereto, or diagonally.

But the two varieties of buildings had, notwithstanding, these other qualities in common; viz., that they were straight-sided, or rectilinear;

—and rectangular also, or nearly so, in the ground-plan of their basements; besides being terminated externally, on all their walls or surfaces, by some of their best and hardest stone or brick.

A direct contrast, therefore, to both the rectilinear and rectangular arrangement, is offered by those antique builders, who,—wandering apparently from land to land,—erected *circles* of great stones; which circles, whether in Palestine, Greece, or Italy, were almost as mysterious to Greeks and Romans, as are to ourselves the similar structures to be seen in France, England, and Scotland. Nor should the arrangement of the class be confined to mere megalithic open-air rings, for its important *circular* characteristic is found equally among the closed-in structures of tumuli, which are not only circular themselves in plan, but are frequently surrounded by either a circular ditch, or a circular bank, or again a circular ring of tall pillar-stones; while the Pyramids of Jeezeh are just as rectilinear and rectangular in the earthen banks that bound them south and west, as in their own masonried bases.

In these 'tumuli,' also, we meet with another radical feature of opposition to the Nileian and

Euphratean examples; for the tumulus has characteristically its only defined and built portions *within*,—while *without*, it ends merely in soft or perishable earth. A tumulus may have been, equally with a pyramid, almost invariably a tomb; but has this constructive distinction, or, that it was fixed chiefly by the place of its centre; while the pyramid was defined by the position of its bounding outside surfaces. A pyramid was thus, as it were, a mathematical definition,—which settles by a distinct circumference, whatever is rigidly *excluded* from its bounds: while a tumulus of antiquarian architecture, is rather a natural-history type, which, by showing the place of the centre, indicates whatever is, by its position, likely to be eminently *included*, within a circle whose size cannot be vouched for.

Or again, a Pyramid is *Promethean*, *i.e.*, careful, well-measured thought *before* the act; while a Tumulus is *Epimethean*, or sudden, thriftless thought *after* the act. The one, in classic fable, brought fire down from heaven, for the benefit of the human race; the other, hid the candle of humanity under a bushel of ills.

Geographical Tracings.

One of the earliest, as it is certainly one of the remarkable, spots on the earth for circular tumuli, —is the neighbourhood of the Gygæan Lake, near Sardis, the capital of Lydia, in Asia Minor.

There are several ages evidently apparent amongst these circular heaps; and one even of the most recent of them is that described in 443 B.C. by Herodotus, as the tomb of King Alyattes, who died about 570 B.C. It is a prince of tumuli, having at present a circular base 3700 feet in circumference, within which a square might be inscribed, sensibly larger than the base of the Great Pyramid of Egypt. Considerable skill too has been shown in levelling the native rock over some parts of that area, and erecting supporting masonry in others, so as to make a platform 60 feet higher than the lowest ground outside.

On that platform a burial chamber seems to have been erected,—and then the whole was covered with layers of earth, rubble, and concrete, in a rounded manner, to a height of 140 feet,—some masonried ornaments being added at the top.

So far, the recent explorations by M. Spiegelthal, the Prussian consul at Smyrna,[1] agree well with the description of Herodotus as to the facts seen and measured by him; but the Halicarnassian's determination of the *age* and *history* of the monument, may be just as erroneous and as much reduced and cooked up to please Grecian vanity, as was his account of the Great Pyramid;—and there is certainly something very similar in both cases, touching his talk about the folly of monarchs and the immorality of damsels of the period,—which more than excites suspicion.

Still, however, both at Sardis, at Ilion, and at Tantalais, on the north shore of the Gulf of Smyrna, are sufficient numbers of these circular-based, central-chambered, tumuli,—to enable Mr. Fergusson to lay down ethnographically that their builders were a Turanian race; and that *that* is why we find no other traces of them than the tombs of their much-revered ancestors; and to say also, that the Smyrniot remains appear as if 'left there most opportunely to authenticate the 'tradition of the Etruscans having sailed from this 'part of Asia Minor, for Italy, about 1200 B.C.'

[1] See *Lydischen Konigsgräben bei Sardes*, by J. F. M. Olfers. Berlin, 1859. 4to, 18 pages, with a map and four plates.

Pelasgic and Etruscan Remains.

Both Pelasgi and Etruscans, equally tomb-building, art-loving peoples, and equally Turanian or ultra-Celtic in origin,—Mr. Fergusson derives from Asia;—and finds them to have migrated in the 13th and 12th centuries B.C., into Greece and Italy respectively: building there often massive city walls, cyclopean gateways, horizontal-stoned arches; but, above all, circular tumuli to their progenitors.

Of these latter constructions, were the so-called treasuries at Mycenæ and Orchomenus. The former, known also as the tomb of Atreus, exhibits the earthen-heap exterior,—an interior bee-hive dome of horizontal stones, and a small rectangular sepulchre hewn out of the rock on one side. The internal chamber likewise betrays the symptom, of having been once lined with plates of bronze,—a very Asian style of decoration;—and still shows the lower half of a column bedizened with zigzags and frequent, thoughtless, inscribed spirals.

The counterpart of this purely Asiatic form, says Mr. Fergusson, may be found in Assyria or Persepolis, but nowhere else in Greece. He notices also, that 'the same (spiral or) scroll ornament 'exists at New Grange in Ireland, in the island 'of Gozo near Malta, and generally wherever 'chambered tumuli are found.'[1]

It was likewise connected in a peculiar manner with the statues of the Syrian goddess, Astarte, or Mylitta; whence Sir Gardner Wilkinson infers, that, from the same volute 'being sculptured 'on the walls of Crendi, in Malta, those singular 'Druidical-shaped ruins' (the Hagar Keem, 'or upright stones') 'are of a people whose religion 'bore some relationship to that of Phœnicia, 'though they are not Phœnician, for the Phœ'nicians would not have made such rude monu'ments.'[2]

The figure of Astarte, however, volute and all, has been discovered in Etruria; and there also, —besides many fine city walls, dearth of temples, but tombs innumerable,—are thousands of circular tumuli; chambered within, a low retaining basement wall without, and a conical earth-heap over the top;—while bronze utensils figure abundantly amongst the discovered contents.[3]

1 Fergusson's *History of Architecture*, vol. i. p. 215.

2 In Rawlinson's *Herodotus*, vol. ii. p. 544.

3 See Canini's *Etruria Antica*, for its fine plates, but, from his

The historic tomb of Porsenna is, indeed, described as made up of 'pyramids,' which one would expect to be square-based and triangular-sided; but the cut given by Mr. Fergusson on p. 266 of his vol. i. shows that other alleged Etruscan pyramids were merely *circular*-based, and moderately *conical*, pillars.

Great Britain and Ireland.

Besides certain examples in the countries between, as at Karnac in France,—circular buildings abound in our own country, witness Stonehenge, Abury, Stenness, and other examples of open-air, erect stones; while New Grange in Ireland, and Maeshowe in Orkney, possess covered tumuli. Sir William Wilde, too, in his remark, that,—'were we to strip the chamber and passage of 'New Grange of the surrounding mound, to re-'move the domed portion of the cave, and to 'replace the outer circle of upright pillar-stones 'at those parts where it is deficient, we should 'have presented to us a monument not unlike 'Stonehenge,'—has given a strong connecting link

Egyptian work, I should say, often doubtful restorations. Not a single compass drawing either is given in any plan of a tumulus, to test the direction of its entrance-passage.

between the two most striking varieties of circular buildings.[1]

Another such connexion is found in the sepulchral character of both; while the tumuli, with their hive-shaped domes,[2] are compared again and again by different authors to the same shaped apartment in the tomb of Atreus at Mycenæ; though all our native examples are lamentably rude, and built in general with entirely unworked stones.

Theories of the dates and builders of these monuments are rife, as every one knows; and various. But, excepting some additions to Stonehenge, evidently of an anomalous order and recent period,—opinions seem now generally converging towards a pre-Christian date and a people from the East. Nor is there much difficulty in imagining these circular-builders of our present chapter,—who fled from Asia Minor in about 1300 B.C., from Greece in 900 B.C., from Etruria in 500 B.C., as Pelasgi and Etrurians, to have been

[1] *Beauties of the Boyne*, by Sir W. R. Wilde, p. 203.

[2] The central chamber at Maeshowe is square in plan, not improbably in consequence of the long natural slabs of stone pаocurable there, and easily forming that shape. But the tumulus outside is conical, the surrounding ditch circular: and the neighbouring stones of Stenness, arranged in a circle. See *Appendix.*

preceded by the Cimmerii, Cymbri, or Cymry, who are traceable also in a similar number of successive habitats, beginning on the shores of the Euxine, and ending in Wales and Ireland,—there is little difficulty, we say, in contemplating the weaker members of these communities driven westward by their foes in a continual migration; losing more and more of their Eastern-made implements and Eastern refinement, and gradually arriving in Western Europe, a degraded, hungry, and forlorn race;—but still keeping up, what Mr. Fergusson maintains is, *par excellence*, the abiding feature of all Turanian origins, viz., the tomb-building tendency for their dead.

Phœnician navigation, and the tin-trade to Britain, may have done a little for this westward movement; but the chief part of it must have depended on the advance by land of large masses of population; hunting, cultivating, or fighting, and frequently occupying for a generation or two as they went. That the Druid priests and many of their religious dogmas of proved Eastern proclivities, were thus derived and descended, is also probable enough.[1] So likewise those puzzles to

[1] See, for a plain and excellent popular digest of the present state

Irish antiquaries and historians of architecture, the *round*-towers of Ireland, standing side by side with a Gothic angular structure, yet more alien than a negro amongst a northern people;—and because, they are a remnant, or are the sole remaining copies, of now extinct originals, once erected in that land by the furthest-travelled westward branches of the *circular* builders.

That these men were not mathematically minded we have already shown (p. 87) by the composition of their chief structures; that they were inclined to barbaric ornament, their platings with metal, and abundant decoration by concentric circles and spirals (fit predecessors of the tattooing of South Sea savages),—sufficiently demonstrate;

of questions touching the 'Ancient Britons,'—two papers in the *Fortnightly Review*, vol. xiv. 1866.

See also, chapter 5, by Dr. Thurnam, in Barnard Davis's and Thurnam's large work, *Crania Britannica*, where the data are gone into at great length and treated with eminent ability and impartiality.

Dr. Stukely's *Stonehenge a British Temple restored to the Druids*, (1740 A.D.) was a book far before its age; and perhaps not yet sufficiently appreciated. His attempt to show by measurement, that such megalithic circles had always been arranged on even and round numbers of the profane Egyptian or Babylonian cubit (20·7 inches in length nearly), and not in feet or any other known northern standard of length,—was both excellent in itself; and though it has been pooh-poohed by more recent antiquaries, I have never heard of any of them having ascertained by actual measure at the place, that the Stukeleian theory would not hold.

and that they were not astronomical, at least in their latter, or British, stage,—the utterly varying azimuthal directions of their entrance-passages to the tumuli, incontestably establish.[1] For it was neither an attempt to be cardinal, as with the Egyptian pyramids; nor diagonal, as with Chaldæan temples; but a free and easy determination, like that of Gallio, to care for none of these things.

[1] Masehowe entrance-passage, from a plate in Mr. Mitchell's volume on Masehowe, enters in from S. 40° W., nearly.

New Grange, similarly enters, according to Governor Pownall, from N. 24° W. And its neighbouring tumulus, Dowth, as marked on the Ordnance Map, from the west; while two smaller neighbours enter from the east.

The tumulus of Alyattes, as given in plan, in M. Olfer's plate 3, has no entrance-passage proper; though there is a modern *forced* passage entering from nearly *south;* and other older ones inside, crossing each other in all directions like the streets of a Saxon town. These efforts have, however, discovered the sepulchral chamber; in a position far from central to the mound, and with an axial direction of North 9° E. The chamber is oblong, rectangular, flat-roofed with stone beams, and reminding one of the King's Chamber in the Great Pyramid; though not so well, or methodically, built, and very much smaller; being in *length*, only 132; *breadth*, 93; and *height*, 82 British inches; against 412, 206, and 230 respectively for the Egyptian room. Outside the Alyattian chamber is a portion of an old passage, exhibiting a curious mixture of rectangular masonry, marble inlayings, and a rounded roof of mere grouted rubble and small pebbles,—which would have horrified the Pyramid-builders of Egypt.

ARCHITECTURAL DATA—*continued.*

CHAPTER 13.

FIRST RESIDUAL ARCHITECTURE.

Hebrew.

THE Hebrews being the most Semitic of all nations descended from Shem,—original and early architecture is not to be expected from them; for the Semites, Mr. Fergusson tells us, never were builders.[1] But then, again, the Hebrews at an early

[1] 'Neither the Phœnicians nor any of the Semitic races were ever 'builders, and we look in vain in Spain or at Carthage, or at Tyre 'or Sidon, for anything to tell us what their architecture may have 'been.'—Fergusson's *History of Architecture*, vol. ii. p. 121.

Again, in his first vol., p. 58, Mr. Fergusson appends a note, which, though it allows in a manner that certain hybrid architectural traces all round the Mediterranean may be 'Phœnician art,'—yet describes it as *copying* only, from other nations; and copying without understanding:—or, as he cruelly adds, but with perhaps, after all, extreme ethnographic acumen touching the recently supposed Semitic origin of Anglo-Saxon peoples—'Something like our 'own architecture, imitating everything, understanding nothing.' Making 'sarcophagi at Tyre, of Egyptian form, but with Phœnician 'inscriptions and for Tyrian kings; Greek ornaments in Syria 'which are not Greek; and Roman architectural details all over 'Northern Africa, which, however, are not Roman.'

period of their patriarchal life went down to Egypt; and were put there through a long and weary apprenticeship to building under a Turanian race; becoming thereby what they would otherwise never have been.

Even then, however, a very distinct architectural place from all other Semites, must be given to the Hebrews,—because in their two earliest examples, the Tabernacle in the wilderness (about 1500 B.C.), and the Temple of Solomon (1012 B.C.)—the designs of those buildings were not their own, but were given by Divine inspiration.

The Bible is most distinct on this point. For, as to the former building or portable structure,—chapter after chapter in Exodus contains the description of the materials, and the forms into which they were to be made up, according to the Divine commands. Commands often expressed in numerical measures; with the repeated solemn injunction also to Moses, 'And look that thou make them 'after their pattern, which was showed thee' (or thou wast caused to see) 'in the mount.'[1] And then comes the conclusion of the work, in the

[1] Exodus xxv. 40; xvi. 30; and Numbers viii. 4. See also, in the New Testament, Acts vii. 44; and Hebrews viii. 5.

words, 'And Moses did look upon all the work, 'and behold, they had done it as the Lord had 'commanded, even so had they done it, and 'Moses blessed them.'

And as to the latter building, we read with equal clearness—

'Then David gave to Solomon his son the pattern of the 'porch, and of the houses thereof, and of the treasuries 'thereof, and of the upper-chambers thereof, and of the 'inner-parlours thereof, and of the place of the mercy-seat,

'And the pattern of all that he had by the Spirit, of the 'courts of the house of the Lord, and of the chambers round 'about.'

'All this, said David, the Lord made me understand in 'writing by his hand upon me, even all the works of this 'pattern.

'And David said to Solomon his son, Be strong, and of 'good courage, and do it: fear not, nor be dismayed; for 'the Lord God, even my God, will be with thee; he will not 'fail thee, nor forsake thee, until thou hast finished all the 'work for the service of the house of the Lord.'[1]

Those works have now all of them disappeared from the face of the earth, as known to man; but their contemporary Biblical descriptions and specifications remain, and will be found of infinite importance in some of the questions which we have to discuss in the latter part of our volume.

[1] 1 Chronicles xxviii. 11, 12, and 19, 20.

Greek.

That Greeks were simply and no other than Greeks, from before the siege of Troy down to the days of Alexander the Great,—even the philosophers of the porticoes in Athens itself, might have strenuously insisted. Yet the durable logic of architectural facts, enables Mr. Fergusson to separate the early days of Pelasgianism, from the totally distinct ones of subsequent Hellenism, or Greek proper ; and even to place several centuries of dark ages between.

Under the former period were built cyclopic walls, circular tumuli, and debased oriental frippery ;—but then came in the Aryans followed by the Dorian invasion ; when the Pelasgi were stricken down, partly compelled to emigrate westward, and partly combined with the conquerors. This mixture of peoples heaved for an unknown time in a new national birth, and then suddenly the architectural result appeared before the world in the temples of Corinth and Athens—temples which had nothing whatever Pelasgic about them ; were Egyptianesque rather in derivation, and were rectilinear, rectangular, and columned in rock-like simplicity and grandeur. So elevated too above foreign Egyptian predecessors, as to be freely allowed the honour of being national and peculiar to Greece ; and worthy also of all that highest of architectural praise, which they have been receiving from the whole of the civilized world ever since.

But this extreme genius in the most permanent of all the fine arts, lasted only from about 660 B.C. to 330 B.C., and passed off as quickly as it came ; for, having been dependent, in the Fergussonian philosophy, on the progress of transition of an Aryan race into a Turanian stock,—the nation subsided into ordinary humanity when the change was completed, and has remained there ever since.

Roman.

What the Pelasgi were to Greece, such were their Etruscan neighbours and predecessors to the Romans. But in an increased degree ; for the Etruscans lasted much longer as an independent people, and modified, almost as a parent, the stripling State that was ultimately to possess the ancient world.

When Rome from 200 B.C. to 400 A.D. had

nearly fulfilled her mission, and become the military tyrant and enriched possessor of all other Mediterranean countries,—then, whatever had been invented elsewhere, and Greek architecture of course amongst the rest, was brought in by wealth and power to contribute to Rome's luxurious enjoyment. But the tastes,—imbued deeply into her inmost feelings during her long foster-nursing by the *circular* builders of the Etruscan cities,—are perpetually cropping out. And hence the liking in Rome for *circular* temples; then, if not the invention of the arch proper, yet its popularization and development; certainly the invention of the Dome, whereof the Pantheon is still the finest example the whole world can show; and, finally, earth-topped tombs.

Canini's restoration of the tomb of Augustus—taken by itself, and measured by our own ideas of cleanliness and propriety,—is inconceivable. A circular arrangement of exquisite porticoes, columns and decorated walls facing to nearly every degree of the circle,—resplendent in all the carving and polish that marble can receive,—and yet with a huge earthen mound on the top, deep enough to grow, and actually growing, trees in!

But that is the very essence of the old Etruscan tumulus; viz., a circular containing wall of masonry below, and a heap of earth above. And if every rain-shower brings down defiling streams of earth over the fair face of the lower wall,—developed though it may have been into exquisite architecture,—why, what is that, other than the well-known Eastern custom of women who weep over a corpse, throwing earth on their heads, and drawing streaks of mud down their cheeks? A building visibly wailing for the dead, with a perpetual sign of lamentation,—so long as a particle of the earthen mound above, remains to be washed down by the rains of heaven, over its rounded cheeks of architectural design below,—such was the Roman devised tomb of Augustus Cæsar.

We strive, and strive in vain, to keep the tombic monuments of our loved ones, fair and bright against the influence of all weather; but the Romans, taking a hint from mound-building Etruscans, used the bad weather itself, in combination with their monument—half of masonry and half of earth,—as a manufacturing arrangement, for keeping up a never-ending appearance of one of the ancient accompaniments of weeping out of respect for the dead.

Only join to such a Romano-Etruscan tomb, a perpetual water-wheel on a neighbouring stream, set to turn for ever a Buddhist cylinder inscribed with prayers;—and the applied mechanics of *quasi*-religious grief, would be complete. But of these *Buddhists* now, we must speak separately.

Buddhist.

Buddhist architecture is essentially circular, domical, tombic, memorial, and ancestral; in so far therefore Turanian and Pelasgic—or of the school of the circular builders,—even to the extent of surrounding their topes, or solid domes, with columnar pillar stones. But, if really descending from some of the south-easterly moving branches of that once central family,—this representative of it soon became, in an Indian climate, wealthy, enervated, given to gaudy display, refined without taste, and delighting in mere profusion and intricacy of commonplace designs.

Some writers will insist on the Buddhist religion, which is still professed by the largest mere numbers of men in the earth, being that of those early patriarchs who were immediately out of the direct Hebrew line; and therefore with some traditional claims to the remembrance, at least, of *ethnic* Divine inspiration. But others attribute its formation to subsequent human enthusiasts, between 600 and 300 B.C.; and much of its literature and so-called history, to times subsequent to the Christian æra.

Be this however as it may, seeing that its earliest architectural remains do not go back further than 250 B.C., nor come down much later than 900 A.D.,—Buddhism can fill no important position in our present inquiry.[1]

[1] Bashan. From a recently published work, entitled, *The Giant Cities of Bashan*, and wherein the author speaks confidently of having seen all, and visited some, of 'the *threescore great cities with 'walls, gates, and brazen bars* of the Rephaim, or giants who dwelt 'in that land both before Moses and before Abraham,'—I had hoped to procure *contemporary* particulars, of those very early times in that interesting part of the world.

But out of seven plates (including the gilt stamp), six represent confessedly *Roman*-columned ruins; while the seventh,—though purporting to exhibit 'a giant's house, with stone ceiling, door, and window-shutter,'—yet betrays a *Roman* colonnade running along the middle of the room, in order to support an original joint in the stone ceiling, where the half-length slabs of which it is composed, meet in the middle!

We can only hope, therefore, that some future explorer may be more successful in arriving at traces of the original architecture, really *contemporary* with the primeval Rephaimites of Syria and the region round about.

ARCHITECTURAL DATA—*continued.*

CHAPTER 14.

SECOND RESIDUAL ARCHITECTURE.

OF all other existing varieties of reputed olden architecture (see the Frontispiece), we read in Mr. Fergusson's carefully prepared volumes,—that

Ceylonese, dates from 161 B.C. to 1235 A.D.;

Jaina in India, from 900 A.D. to 1580 A.D.;

Southern Hindoo, with its so-called 'Gopuras,' or 'Gate Pyramids,' from 800 A.D. to 1600 A.D.;

Northern Hindoo from 600 A.D. to 1700 A.D.; and

Cambodian,—recently discovered and much admired for its superior taste to ordinary Indian and Malayan productions, from 980 A.D. to 1500 A.D.; (being the work of an emigration from Cashmere, driven thence by the early Mohammedan emperors,—and carrying with them many reminiscences of the later Greek art from Bactria and Alexander the Great's other central Asian conquests).

Chinese architecture should afford ancient examples, as the people themselves have been permanent inhabitants of the same land from the very earliest times. But, whether from their building materials being bad, their weather unpropitious and destructive, or European exploration still in its infancy,—nothing is known of that land's structures earlier than 240 B.C., the date of 'the great wall.'

While finally, the *Mexican* and other *American* built remains,—including some erroneously named *pyramids*,[1]—but much more like the terraced,

[1] In a recent publication, such undue prominence has been given amongst true and well built stone pyramids, to the mere terraced mound of Cholula in Mexico,—as being 'a pyramid' which has 'about twice the length of basis of the Great Pyramid of Gizeh,'—that it may be worth while to quote, from Mr. Fergusson, that if the Cholula mass is really so very large in the base, it is less than half the height of the Great Pyramid; is broad and flat on the top, (where it was intended to support a hollow temple,) 'is composed 'of badly burnt bricks and mud, and is now so overgrown with 'trees, that it is difficult to make out its form.'

America has also mounds of pure earth, which are occasionally older than her built remains. These mounds are often round, but sometimes in the shape of a ground plan of snakes, tortoises, and other animals. Sufficiently original therefore, but neither equal in age, to many of the works of the circular builders in Europe and Asia; nor similar to them in possessing, as a rule, internal chambers and entrance-passages, whether sepulchral or otherwise.

stair-cased, and temple-bearing, structures of Mesopotamia, than *the burial pyramids* generally, or the Great Pyramid in particular, of Lower Egypt, —date only from 500 A.D. to about 1500 A.D.[1]

These, therefore, close the list of all the remaining known architecture that we need appeal to, for assisting our special inquiry into primal times.

[1] With reference to the oft-mooted question of the origin of American architecture, and its aboriginal inhabitants as well;—the following particulars were recently communicated to me, by Mr. George Davidson; Astronomer to the United States Coast Survey, —and just returned from an inspection of the Coast of Alaska:—

'The whole north-west coast is peopled by the same race of 'Indians that extends certainly to the northern boundary of California. North of Alaska Peninsula are the Esquimaux. But the 'whole of the Aleutian Islands, part of Alaska Peninsula, and the 'Archipelago of Kadiak, are inhabited by the "*Aleutians*," a 'different race from either, and partaking in many characteristics 'and features with the Japanese. It is a fruitful theme for investigation, and I intend to pursue it when on the Pacific this 'spring. The Kamtschatka branch of the Great Japanese Warm 'stream runs into Behring's Sea on the western verge of the Aleutian 'Islands, and in 1862 (Sept.) a Japanese junk, carried by this stream 'from Japan, was wrecked on Alton, the westernmost of the 'Aleutians. They were rescued in 1863. The main stream has 'carried Japanese junks to the coast of Oregon and Washington so 'late as 1833.'

For *plates* of the American mounds and pyramidal-based temples, together with similar constructions in Europe, Scythia, and Africa, see a paper read by Mr. St. John Vincent Day, C.E., in February 1868, before the Philosophical Society of Glasgow, upon *Some points in certain Theories concerning the Purpose and Primal Condition of the Great Pyramid of Jeezeh.*

CHAPTER 15.

ARCHITECTURAL RESULTS.

IF we should now,—after our rather extensive review of all the oldest known remains of architectural skill in any and all the countries of the earth,—resume, with increased powers, our first investigation touching both the recently alleged invariable pyramidal form, and one constant method of astronomical emplacement, asserted as common to all primal structures;—an unexpected answer comes forth.

For, on making such examination, we immediately find, not only that there is amongst all known architectural works, only one territorial instance —viz., the Lower Egyptian,—of true and pure geometrical Pyramids; but that it is only there, that the asserted astronomical orientation has any place.

Elsewhere, either no astronomical intention

whatever appears, as at Thebes and Meroe, and with the circular buildings everywhere ; or, as in the Chaldæan, Assyrian, and Babylonian structures, an opposite orientation law crops out ; *i.e.*, in the basement walls there being directed not on, or towards, but away from, the cardinal points ; and to the utmost possible extent.

This very defined and signal fact of astronomical opposition in Euphratean practice to Nileian precedent,—is further shown, by a consideration of all the data, to have been most intentional, settled, and determined on the part of every branch and age of the Interamnian population. For, however the various successive empires there rose or fell, and whether their remains are first met with on the sites of the great burial cities near the Persian Gulf ; or, whether afterwards they are found coming out in splendour in the cold north-western regions of the Assyrian empire, and its severer metropolis of Nineveh ; or after that again flourishing once more in the burning south, within the mighty ramparts of gaudily-decked, and almost tropically-heated, Babylon the magnificent ; and whether the tastes of the local sovereigns and their people ran now on temples to their gods, and now on palaces for their kings,—still that determination to hold on to a peculiar, and scientifically a most inconvenient, orientation,—is predominant in almost every case.

Something also of a *growing* or accumulating and strengthening feeling was this in Mesopotamia, being realized on a larger and larger scale on every successive occasion; so that while in Egypt we discover the best and grandest of its Pyramids at the beginning of its series (see Frontispiece),—in Mesopotamia, on the other hand, the largest and most typical of all its peculiar terrace-based temples, is the latest of the whole. In everything, therefore (except where the circular builders differ from both, and show so little care for any species of science, that science can hardly include them in her intellectual ranks), the Euphrates series of buildings, though nearly parallel with, but not quite so ancient as, the series of the Lower Nile,—is clearly their very opposite in both manner, arrangement, external appearance, internal intention, and even whole history of birth, invention, and development.

Now something so marked as this, and when

we had been led by literary men to expect close uniformity,—makes it well worth our while to inquire, whether there be not some other data which can be referred to for additional testimony; something equally *contemporary* with the architectural, perhaps older, certainly varying the point of view, and otherwise also proper to be introduced into the present inquiry.

ARCHÆOLOGICAL DATA.

CHAPTER 16.

FLINTS OF THE DRIFT.

'WHILE we have been straining our eyes to 'the East,' says Sir John Lubbock,[1] 'and eagerly 'watching excavations in Egypt and Assyria, 'suddenly a new light has arisen in the midst of 'us; and the oldest relics of man yet discovered 'have occurred, not among the ruins of Nineveh 'or Heliopolis, not on the sandy plains of the 'Nile or the Euphrates, but in the pleasant val- 'leys of England and France, along the banks of 'the Seine and the Somme, the Thames and the 'Waveney.'

The supposed new light thus eloquently referred to, was that lit in 1846, when M. Boucher de Perthes published his first account of certain flint weapons, found in special gravel, and drift, beds

[1] *Prehistoric Times*, by Sir John Lubbock, F.R.S., p. 268.

near Abbeville in France; and he was soon followed, imitated, and confirmed, by half the geological world.

These drift-beds have been estimated by geological processes, to be 30,000, 50,000, or any greater number of years old; and contain the bones of both the mammoth and sundry other now extinct animals (forming therefore a post-pliocene, or post-tertiary geological formation); and though no bones of *man* are findable therein, yet hundreds and even thousands of flints, chipped with an evident intentional mental purpose, have been collected,—and are held by those most skilful in such inquiries to testify to a human origin.

Man, then, argues this new school, must have lived when these gravel-beds were being deposited; when the mammoth, the woolly rhinoceros, the cave-bear, and divers other now extinct animals, roamed over the earth. Man's own bones have disappeared, probably because they were small, especially seeing that no bones of animals smaller than man, have come down to us in these gravels; but man's lasting weapons of durable flint, still remain to bear testimony both to his skill in manufacturing them, and his contemporary existence with a Fauna and Flora essentially different from what now prevails over Europe.

In the midst of all the ecstasies of delight, with which these very recent discoveries have been hailed, there has perhaps not been sufficient remembrance, that so far back as 1824, the late Rev. Dr. Fleming[1] maintained the co-existence of man with the above-mentioned now extinct geological animals.[2]

Also that their remains occurring only in superficial strata, are connected, according to him, with the last, or modern epoch only, of the earth's history; and that their, the animals', final extinction was greatly accelerated, or even entirely brought about, by human hunting. Hunting—indicates the learned minister—formed the inveterate occupation of men, from the earliest days; and continued in full force so long as there was anything

[1] In Jameson's *Edinburgh Philosophical Journal*, vol. xi., 1824; and vol. xiv., 1826.

[2] Vol. xi. p. 303.—'(5.) The remains of those extinct animals 'occur only in the superficial strata, and in fresh-water gravel or 'clay, and may be viewed as connected with the last or modern 'epoch of the earth's history. (6.) Man was an inhabitant of this 'country at the time when these animals' (mammoth, elk, rhinoceros, hippopotamus, cave-bear, hyæna, etc.), 'now extinct, flourished, 'his bones and his instruments having been found in similar situa-'tions with the remains.'—Rev. Dr. FLEMING, 1824.

well worth their hunting: and he even considers the process to be so sharp and powerful as an exterminator, that he apparently sees no reason why all the animals known to have been at any time contemporary with man, but now extinct, should not have died, or been killed out, within the last 6000 years.

Some animals, and generally the larger, are exterminated much more quickly than others. So that, as we know by our own personal experience at the Cape of Good Hope,—the lion, elephant, rhinoceros, eland, and many other large animals, have entirely and recently vanished out of a tract of country far larger than all Great Britain,—while the leopard, hyæna, jackal, duiker, steinbok, etc., remain. Only two hundred years ago, the extreme southern point of Africa, was a fit representation of the great mammal ages of geology—from the number and variety of huge and ponderous beasts that roamed over, and grew so astonishingly large, upon its plains of scanty verdure. Two centuries however, assisted no doubt by fire-arms and the migration of a civilized race, have sufficed to produce there a zoological change, akin to those which are generally attributed to uncountable numbers of years, by the pure geologist.[1]

No one, however, was more instant in season and out of season than Dr. Fleming, in pressing upon the attention of the learned world, the rapid manner in which divers birds and animals have been thinned down in their numbers, within the period of Saxon history, and thereby brought to the verge of geological *extinct* species; and yet even he was not able fully to realize the rapidity of the process. For while we have all been taught, almost to shed tears over the fate of the Dodo, which no man on earth shall ever see again,—and to inveigh against the unscrupulous destruction of that strange-looking fowl by early southern navigators;—the 'great auk,' we are now told, has died out of existence within the present century, and amongst ourselves; and Dr. Fleming himself, had, in a tamed condition, the very last known member of the race;—though, apparently without being fully aware at the time, that the

[1] For a plain, unvarnished statement of these great animalian facts, as they existed so short a time ago, in the neighbourhood of what is now the peaceful city of Cape Town,—see 'The Record; or 'a Series of Official Papers of the Dutch Government, from 1649 'to 1720 A.D.,' published in Cape Town in 1838, under the editorship of Lieut. D. Moodie, R.N.

poor individual bird before him, was in truth the carrier of such a weight of destiny almost accomplished.

Dr. Fleming, then, was exceedingly inclined towards short dates for the contemporaneousness of man with—not the whole duration of the species of the mammoth,—but merely, some of the later members of it. And perhaps he would have retained those short dates, even though he had lived down to the present day,—when almost every one is demanding more *time*, wherein to place in order the real or supposed facts of the early history of the earth.

Mr. W. Carruthers has no doubt pointed out,[1] that there is a permanent and characteristic difference between all zoologists on one hand, and geologists on the other, as to the ages which they assign to the same natural and primeval facts: the former men of science being inclined to comparatively small, and the latter to enormously great, periods of time. But then, so many other natural-history students have recently gone along with the popular geological current of the day,

[1] 'The Antiquity of Man,' in vol. xvi. of *British and Foreign Evangelical Review*, p. 383, etc.

that we cannot be certain of what the minister of Flisk's opinion would have been *now*, in presence of the later facts and papers, before which so many of his friends have succumbed. Indeed we rather suspect, if the mean judgment could be taken of all existing natural historians, zoologists, and others,—and especially those selected examples of them filling places of honour in metropolitan scientific societies,—it would undoubtedly be found to attribute to the rough 'flint hatchets, knives, and spear-heads' of the drift —a vastly superior age to that of any of the architectural monuments we have yet reviewed.

Such an age, indeed, that, if some of the earlier of these latter objects date from 4000 years ago, —the flint implements of the drift, say these gentlemen, must date from 40,000 or 80,000 years since, or more. Whether these scientists are right, or wrong, in this view, we do not now pause to inquire,—but shall only examine the case, on its own advocates' showing, in order to ascertain what may be its best and most favoured claims to enter into our present investigation.

Chipped Flints.

We should observe, in the first place, that the said flint weapons being merely chipped into shape,—and not ground on the edges or polished over the surfaces, as with most of the stone weapons of known savage tribes in modern times, belong,—according to those who make savage manners and customs their study in the present day,—to the very lowest conceivable stage of savagedom. And this main fact, combined with the inferred vast age of the specimens, is thus employed by Sir Charles Lyell in throwing a most powerful light on the nature and time of origination of *intellectual* man (our very title and subject), on the surface of the earth. We give the whole of Sir Charles's long paragraph from end to end,[1] so that there may be no mistake in gathering his real meaning :—

'It has sometimes happened that one nation has been con-'quered by another less civilized though more warlike, or 'that, during social and political revolutions, people have 'retrograded in knowledge. In such cases, the traditions of 'earlier ages, or of some higher and more educated caste

[1] *Antiquity of Man*, by Sir Charles Lyell, third edition, revised, pp. 378 and 379.

'which has been destroyed, may give rise to the notion of 'degeneracy from a primal state of superior intelligence, or 'of science supernaturally communicated. But had the original stock of mankind been really endowed with such superior 'intellectual powers, and with inspired knowledge, and had 'possessed the same improvable nature as their posterity, the 'point of advancement which they would have reached ere 'this would have been immeasurably higher. We cannot 'ascertain at present the limits, whether of the beginning or 'the end, of the first stone period, when Man co-existed with 'the extinct mammalia, but that it was of great duration we 'cannot doubt. During those ages there would have been 'time for progress of which we can scarcely form a conception, and very different would have been the character of the 'works of art which we should now be endeavouring to interpret,—those relics which we are now disinterring from the old 'gravel-pits of St. Acheul, or from the Liége caves. In them, 'or in the upraised bed of the Mediterranean, on the south 'coast of Sardinia, instead of the rudest pottery or flint tools, 'so irregular in form as to cause the unpractised eye to 'doubt whether they afford unmistakable evidence of design, 'we should now be finding sculptured forms, surpassing in 'beauty the masterpieces of Phidias or Praxiteles; lines of 'buried railway or electric telegraphs, from which the best 'engineers of our day might gain invaluable hints; astronomical instruments and microscopes of more advanced construction than any known in Europe, and other indications 'of perfection in the arts and sciences, such as the nineteenth 'century has not yet witnessed. Still further would the 'triumphs of inventive genius be found to have been carried, 'when the later deposits, now assigned to the ages of bronze 'and iron, were formed. Vainly should we be straining our 'imaginations to guess the possible uses and meaning of 'such relics—machines perhaps for navigating the air or 'exploring the depths of the ocean, or for calculating arith-'metical problems, beyond the wants or even the conception 'of living mathematicians.'

Now after reading this most important judgment by one of the first geologists of our age, and apprehending fully his grounds of argument,—it will appear clearly, that his total conclusion will only be intensified, by any further discoveries or discussions, which may either prove or imply, a very much greater age still, for the flint implements dug out of the drift, than what we have yet recorded. So that if any particularly go-a-head geologist shall draw still more largely on those stores of untold time, which cost him and his brethren so very little,—and put millions and even billions of years where we have put thousands only,—that will but make it infinitely more certain than ever, according to the involuntary testimony of Sir C. Lyell,—that man, at that excessively removed period of time, could *not* by any possibility have been *then* 'endowed 'with superior intellectual powers, science super-'naturally communicated, inspired knowledge, and 'the present *improvable* nature of human kind.'

In fact, though whole shiploads of chipped flints may be dug out of any number of gravel-beds, and may also be claimed by anthropologists as *contemporary* remains of those who shaped them ages ago,—they testify nothing as to *intellectual* man; and therefore have little more to do with the subject of our investigation, than the mere gravel itself.

What we alluded to under our title as *intellectual* man, and meant to imply thereby, was,—man as he is now, or with the mind wherewith he was already endued in the days of Job, of Abraham, and of Moses;—and that state which certain eminent anatomists are inclined to distinguish, within the limits of their science,—as implying an organized creature with an immensely preponderating amount of *brain,* to that possessed by all other known living creatures; or again, as a being with an almost unlimited capacity for *mental improvement.*

This last feature or form of the statement, is evidently a part of Sir Charles Lyell's means of distinction; but when he adds thereto, 'science 'supernaturally communicated, and inspired know-'ledge,' though only as mere traditions and notions that may have originated out of the accidents of human history,—it is evident that he brings on the field the teachings of religion, with regard to the Divine creation of man. The teachings of many systems too of religion,—though he does not point out anything better, or essentially different, in one, than another.[1]

Yet inasmuch as the Egyptian and Assyrian religions have given dates of several tens, or even hundreds, of thousands of years for man living on the earth under kings, princes, and civilized rule very much as at present; and one of the dates for 'Ikswaku and Buddha' of the Indian religion, goes back so far as to 2,183,102 years B.C.,[2]—it is perfectly impossible to square such dates, with the most important integration which Sir Charles Lyell has performed, upon all that we know, by geology and otherwise, of former and present works of man. Those Gentile dates therefore,

[1] 'The inventors of useful arts, the poets and prophets of the 'early stages of a nation's growth, the promulgators of new systems 'of religion, ethics, and philosophy, or of new codes of laws, have 'often been looked upon as messengers from Heaven, and after their 'death have had divine honours paid to them, while fabulous tales 'have been told of the prodigies which accompanied their birth.'—P. 504, 3d edition of Lyell's *Antiquity of Man.*

[2] *Useful Tables of Indian Metrology and Chronology*, by James Prinsep, Royal Asiatic Society, Calcutta.

according to the general tenor of his whole paragraph, cannot be true; or the character of geological findings would be totally diverse from what they are. Wherefore it is well to remember, that the Revealed religion of the Bible tells a very different story, as to the true and limited era of man's early civilization.

Several worthy men,—indeed Dr. Prichard, in his ethnological researches, amongst others,—have confessed, that the earlier chapters of Genesis hardly seem to them capable of a very definite interpretation as to the exact chronology of man; —though they all agree that the Bible date of the *Divine inspiration* of man, or, when 'man became a living soul,'—the important point in a work on religion,—cannot be put further off from our times, on any rendering of the Biblical text, than between 6000 and 7000 years ago.

Until, therefore, *intellectual* monuments equal to those of the present day, or to those much more advanced constructions which Sir Charles Lyell has sketched out,—shall be discovered, nature placed in drift, gravel, or clay-beds, which geologists shall testify or prove to be older than 7000 years,—there will have been nothing yet discovered by geology to impugn the Bible date of *the Inspiration of man.*

Almost needless is it to say, when our readers can examine for themselves, as well as we can, both Sir C. Lyell's work on the *Antiquity of Man,* and Sir John Lubbock's on *Prehistoric Times,*—that neither of these very comprehensive works on recent geological discoveries, contains the slightest allusion to any such high intellectuality characterizing the works of any times, which they say, are more, more in any way, certainly not much more, than 6000 years, or even 4000 years old: certain possibilities in Egypt being excepted, and which we shall discuss in chapters 18, 19, and 20.

Yet what neither of those great leaders in their respective subjects has ventured to hint or even hope for,—has very lately been asserted elsewhere so roundly, repeatedly, and positively—both in print, and in speech before scientific societies—that we must devote the next chapter to its careful consideration.

ARCHÆOLOGICAL DATA—*continued.*

CHAPTER 17.

PYRAMIDS OF EUROPE.

MEETING unexpectedly, in a work published last year,[1] with the striking phrase,—'*those magnificent though rude western Pyramids that constitute the grand old mausolea of Ireland and Brittany,*'—I could not but wonder how Mr. Fergusson had omitted all mention of such apparently important monuments, from the otherwise comprehensive pages of his *Architectural History!*

Very soon too, another recent work was found[2] speaking, not only freely and confidently of the 'Boyne Pyramids,' but giving to one of them, at present known by the name of New Grange, the following glorious title of praise :—

[1] *Archaic Sculpturings of Cups, Circles, etc.*, by Sir James Y. Simpson, Bart. and M.D.; 1867; p. 115.

[2] *Beauties of the Boyne*, by Sir William R. Wilde, F.R.I.C.S. and M.R.I.A., 1849. Pp. 202, 203, 224.

'The most splendid specimen of this description, which we 'know of, in Central or North-Western Europe, is the magni-'ficent mausoleum at New Grange, which may well be deno-'minated *the Great Pyramid of the West*.'

Thus much for its architectural character; and then for its *age*, the same author writes of New Grange, 'it may be coeval or even anterior to its 'brethren on the Nile.' So that the Scriptural indications of Shinar and Egypt being the oldest scenes of man's architectural triumphs, would seem to require alteration; and the name of Ireland should be read in their place.

Some other localities have also much glory, both architectural and intellectual, claimed in their favour. For, of the Brittany so-called 'pyramids' it has recently been stated,[1] that 'they are 'perhaps older than those of Egypt, for they con-'tain implements and weapons only of stone,—'not of bronze or iron. They were reared in the 'stone age,—the Egyptian tombs were reared 'after metallic tools were invented.'

While with still greater boldness was it said on the same occasion,—'But if not content with the 'age of these enormous western tombs or pyra-

[1] At a meeting of the Royal Society, Edinburgh, December 16, 1867; see Appendix 2.

'mids, we had, besides cromlechs and megalithic 'circles, other remains of intellectual man in 'Europe undoubtedly older than the oldest Egyp-'tian Pyramid.' And a piece of carved ivory, described as 'that accurate piece of drawing or 'sculpture,' was then alluded to, as being 'far, 'far older than the age of the oldest Pyramid.'

But even these claims of enormous antiquity for intellectual man, in this north-western part of the world, are overpassed by certain others made here also for certain sculptured stones (some of them similar to, if not also identical with, many in the interior of New Grange); and whose age their admirers mostly express in a series of dim and misty allusions to archæological periods of bronze and of stone almost interminable, having intervened since they were executed; as well as to huge physical changes, such as the time 'when 'possibly England was still geologically united to 'the Continent, and the Thames was only a tri-'butary of the Rhine;'[1] ages and changes these last, for which geologists usually demand tens, and even hundreds, of thousands of years.

Here then, we have positively,—though in the

[1] *Archaic Sculpturings*, p. 145; by Sir James Y. Simpson.

last chapter we had not,—modern science, as practised by some persons, in direct, and unapologized for, antagonism to Revelation. More completely so too, the further the case is examined into,—for these large assertions of archæological time for anything bearing, or supposed to bear, a highly intellectual character, are almost invariably found to underlie still larger demands, by the same persons, for comparatively unintellectual remains. For these, they claim indeed the most enormous periods; but during which they argue that man may have been developing or improving himself by slow degrees, out of a primal state of bestial ignorance and simplicity—right up to his highest condition in modern times, without having either required, or received, any inspiration or other supernatural aid from his Creator.

Now with the general question of whether these writers are religiously right or wrong,—whose minds find such pleasure in pushing off into the remotest distance of time all symptoms of the direct action of a personal God, and denying every trace of the Scriptural Deluge, on the affairs of all human kind,—we have not here to do. But we cannot avoid taking scientific note, that both an egregious difference of opinion from the Biblical narrative, and an astonishing statement of absolute fact, in the matter of what countries first were inhabited by intellectual man, and when,—has been brought up before us. Brought up too in a manner, and by parties, we cannot ignore; and maintained not a little upon the strength of many asserted qualities and characteristics of the so-called Irish, as containing the finest examples of all the European, Pyramids, and some of their component stones. There is no other course therefore possible to us, as mere searchers after scientific truth, than to examine what are the observations, examinations, or arguments, upon which the existence of the said most remarkable qualities and characteristics of these structures, is asserted.

Tested by Geometry.

We have already shown, p. 13, and partly repeat below,[1] the strict geometrical definition of a

[1] (1.) A Pyramid is a solid whose base is a regular rectilinear plane figure, and whose sides are plane triangles, having all their vertices meeting together in a point above the base, called the *vertex* of the Pyramid.

(2.) Pyramids take particular names from the figure of their base, as triangular, square, rectangular, pentagonal, etc., etc.—*Ordinary Geometrical Definitions.*

Pyramid. How near thereto, then, come the Irish Pyramids?

In Sir William Wilde's woodcut view of New Grange,—the chief of these monuments,[1]—we see merely a low, broad, straggling, mound, partly overgrown with a hazel plantation, and sprawling over a field whose surface dips exceedingly towards one side. But as this is the present state only, and may be very different from the building's original condition, we refer to the more methodical account by Governor Pownall, in the second volume of the *Archæologia*, date 1770: a description extending through forty quarto pages, discussing several previous accounts, and fortified with measures expressly taken for the Governor, by a Mr. Samuel Bovie, a land-surveyor: these measures too, being the latest we can hear of.

Now in his Plate 20, where Governor Pownall gives both a plan (to be taken in connexion with the explanatory letterpress on p. 253), and a restored sectional elevation,—it appears that the base of New Grange was—in ground plan—not a rectilinear figure of any kind, but *circular;* and its top, not pointed, but *broad* and *flat;* the attached

[1] *Beauties of the Boyne*, p. 189.

scale giving a height from the lowest part of the base = 67 feet;[1] with a horizontal breadth at top of monument = 118, and at bottom, of some 319 feet.

All this is evidently the very antithesis of a Pyramid; and while the circular base receives a special confirmation from the remains of a concentric and environing ring of standing pillar-stones still to be seen,—the broad flat top testifies to its having been an *original* feature, by showing a similar large pillar-stone upon it, in the time of Mr. Lhwyd, the first post-mediæval describer.[2]

Now such a sort of lithic top-knot existed on the summit of the circular tumulus of Alyattes, in Lydia,[3] and on most of the Etruscan tumuli; —while, as for the outside *circle* of stones surrounding the base, that is exactly borne out by

[1] The Surveyor's strict measure gave only 42 feet for height, and it is not quite clear that Mr. Pownall was justified in altering his figures.—See p. 252 of *Archæologia*, vol. ii.

[2] Much material had been removed in previous years, for mending roads, but Mr. Pownall shows that it was taken from the side, and not *the top of the* monument.

[3] 'The upper portion is sand and gravel, apparently brought from 'the bed of the Hermus.' On the summit are, 'the remains of a 'foundation nearly 18 feet square, on the north of which is a huge 'circular stone, 10 feet in diameter, evidently placed there as an 'ornament, on the apex of the tumulus.'—Description of the tomb of Alyattes, in *Researches in Asia Minor*, by W. J. Hamilton, 1842; vol. i. p. 145.

another and undoubted tumulus or *barrow* in Arcadia, described of old by Pausanias; and thus translated by Pownall,—'I contemplated 'the tomb of Æpytus with a studious and curious 'reverence, because Homer mentions it in so 'marked a point of view. It is a tumulus[1] of 'earth of no great size, surrounded at the foot or 'base with a circle of stones.'

Hence New Grange, though said to be composed of loose pebble stones chiefly, rather than mere soft earth,—but yet having enough of such earth and garden soil, as to be able to grow a flourishing plantation of hazels on its surface,—can evidently, by virtue of its shape, only claim the name of tumulus, heap, hillock, or *moat*,—as indeed the Ordnance Survey both defines and expresses it,[2]—and is a work of the circular, not the Pyramid, builders. So too, seem to be, from the best descriptions yet published, all the other lesser archæological forms, heaps, or barrows in its neighbourhood.

[1] The word used by the old Greek itinerant is not *πυραμίς*, which he would surely have known how to apply,—but χῶμα; that is, *agger* or *tumulus*, in Latin; and *heap* or *hillock*, in English.

[2] See Maps 19 and 20, of Ordnance 6-inch scale, for county of Meath.

Not one of these either is, or ever was, a 'Pyramid,' in shape, or according to the geometrical definition thereof; or has, or ever had, the least right to the term.

With most persons this would settle the Irish Pyramid question at once and completely; not so however with two, in particular, of the archæologists already mentioned; for they seem to have decided for themselves and a few followers, that the word 'Pyramid' means a species of sepulchre; and also that such sepulchre, if it affects one external form rather than another, is more likely to be *rounded,*[1] than what all the rest of the world would call, pyramidal.

But though a sepulchre be undoubtedly one of the uses to which the Egyptian Pyramids were put,—the name is most assuredly not applied there to sepulchres in general; nor in fact to anything else than angular masses of masonry built in the well recognised external form, or

[1] 'First and most notable is a dome-roofed stone chamber, containing the remains of one or more bodies, and approached by a 'covered way, the whole being enclosed in *a large earthen tumulus* 'or *barrow*, and generally surrounded by a *circle* of upright pillar-'stones. *This is the true Pyramid*, modify it as we please, of which 'the type is to be found in those great oriental monuments, with 'the characters of which all are acquainted'!—*Beauties of the Boyne*, by Sir W. R. Wilde, p. 224.

closely approximating thereto, of mathematical Pyramids;[1] and this, whether they contain internal tombs or not. That such tombs are not held to be necessary there to the name of a Pyramid, is well seen in the case of the 'second Pyramid' of Jeezeh; which, though it was thought, before its opening by Belzoni, to have no chambers at all in it, and to be an entirely solid mass of building—was yet called by all nations a Pyramid, just as well as the known chambered and tombic structures near it.

More extensively still,—what are called by divers authors the 'Assyrian Pyramids,' are merely somewhat pyramidally shaped masses of brick or masonry; always supporting on the top, a hollow temple for living worshippers to attend at, but seldom known to contain sepulchral chambers, with entrance-passages, underneath.

The same precisely may be said of the so-called Mexican, and other American, Pyramids (Fergusson's History, vol. ii. pp. 764-768); for the chief

[1] That some of the Egyptian Pyramids, having been badly built and of poor materials—are now, after 3500 years' exposure, reduced to mere rounded heaps of rubbish,—is evidently not against our argument; for that bases itself entirely on what the original builders intended to make, and did succeed in making, whether in Egypt or Ireland, or elsewhere, in their own time.

known use of the pyramidically shaped part of the structure there, was to act as an elevated base, to give dignity to a temple raised on the top; and to allow the religious ceremonies performed thereat, to be witnessed by crowds of people in the plains below. Wherefore, even as regards the *popular* use of the word 'Pyramid,' apart from its geometric definition, the name has been most abundantly employed by Anglo-Saxons both in the New and Old Worlds, to indicate a peculiar *external* shape, and not some one possible *internal* use, of many and many a building, or other mass carved out, or made up, of solid and dense materials.

Nor is this great fact altogether unknown to, and unadmitted at times by, Sir W. Wilde, and we suspect most of his school as well; for, in the former's work already referred to, there is an extract quoted with apparent approbation from Dr. Molyneux, 1699 A.D.,—as to there having been found lying on the floor of the inner chamber of New Grange, when it was first entered in recent times, 'a slender quarry stone, five or six 'feet long, shaped like a Pyramid.'[1]

[1] *Beauties of the Boyne*, by Sir W. Wilde, p. 202.

The little band of archæologists just mentioned, may indeed assert thereon, that *they* hold, that said stone was therefore shaped like a hemispherical tumulus, or heap, of soft earth just emptied out of a bushel; and had a little sepulchral chamber and entrance-passage worked inside it. But others believe, that the common sense of the world will be dead against such an idea; and will hold Mr. Fergusson perfectly blameless for not having put New Grange's *circular barrow* at the head of all *triangular-sided Pyramids*, in his careful History of Architectural Forms.

Tried by Astronomy.

Yet New Grange and its fellows need not lose any of their alleged high intellectual character, merely on being called by another name than that of 'Pyramid,'—if solid proofs of said high and noble mental qualities are really there. But in that case, where precisely are they; or, in what do they consist?

They are not in external shape,—for there is no science in imitating the form of an earth heap, and with earth materials—which easily dispose themselves in that shape, when thrown together almost anyhow. But then New Grange has in the confession of all men a sepulchral chamber, and that chamber an 'entrance-passage,' with the astronomical direction of 'very nearly north and south;'[1] and that is precisely the ground of much astronomical praise of the Great Pyramid and *its* entrance-passage.

True; and at the Great Pyramid, the ordinary popular term of 'very nearly,' is found by careful measure to mean something less than one-thirteenth part of a degree, from due north and south. But how much does the same phrase imply at New Grange?

On turning to Governor Pownall's treatise, we find 24°!

While this quantity too, is given as being to the north-west,—the entrance-passage of another of the Boyne heaps, is said to open out to the west, and another to the east.

Proved by Mechanics.

If not in its astronomy, then, does the intellectual force of New Grange reside perhaps in its mechanical style of structure and composition?

[1] *Beauties of the Boyne*, by Sir W. Wilde, p. 194.

To this end, we may well examine Plate 21, p. 254 of Governor Pownall, and the several woodcuts of Sir W. Wilde's most effectively and charmingly illustrated volume. The former gentleman is frequent in finding constructive resemblances between New Grange and the well-built Great Pyramid of Egypt; and at his page 268, notices—that a few feet within the entrance of the latter, the passage is reduced to so narrow a strait, that travellers can only creep, or be pulled through, lying down at full length, prostrate and serpent-like.[1] Now there is a similar strait, says Governor Pownall, just within the entrance of New Grange; and though he first thought on looking at it, that it was caused by some of the stones lining said entrance having tumbled over,—yet hearing of so similar a strait at the Great Pyramid, he concluded that it must have been intentional and important: and that the same architectural idea had governed both.

Yet though that strait or narrow part of the entrance-passage did so long exist to trouble visitors at the Great Pyramid,—it was artificially produced *within* the passage wall by mere impor-

[1] See also *Life and Work at the Great Pyramid*, vol. i. p. 370, by C. Piazzi Smyth.

tation of sand and stones from without, by modern Arabs ;—and these adventitious and gratuitous obstructions having been at last removed by Colonel Howard Vyse,—behold the real entrance-passage of the Great Pyramid, a long, rectilinear, rectangular tube, of well prepared, truly planed, and exquisitely close-jointed masonry,—admitting of measurements for breadth and height to tenths and even hundredths of an inch ; and with no approach to any narrowed or *strait* part in it at all. The difficulty then vanishes at Jeezeh. But how about the other strait at New Grange, supposed to have been made by a similar mental intention of the original builders ? Alas ! it seems in so far a real case of that entrance-passage itself,—never very truly rectilinear, and with many ins and outs in its course—becoming much worse at one point, through the very bad masonry there altogether giving way.

Some of the stones of New Grange gallery, are big enough to require much manual force to have moved them to their present positions. But, rude to an inconceivable degree, unworked, indeed untouched by any instruments of the mason,—though scrawled over with the engraved tattooing marks of the savage,—even the best of these stones lining the passage, and forming the lower walls, of the central beehive-shaped chamber of New Grange, are mere mis-shapen, natural pillars or slabs;—placed in wriggling, uneven fashion, and filled in between, more or less, with loose, uncemented rubble-work of small boulders.[1]

The sight is distressing to any one who has the smallest idea of using the rule and the level, or of practically realizing any of the mathematical solids or plane surfaces. One and only one representation speaks of something better; it is the picture of a diagonal-brace ornament on a stone over the doorway—as given by Sir J. Y. Simpson, in the Plate 29, fig. 5 of his *Archaic Sculpturings.* I had looked and looked again at that engraving, feeling that there was indeed therein a proof of higher mechanical knowledge and technical education, than what all the other reproductions throughout that well-filled work, of early British and Irish art, display. But, alas again! Sir W. Wilde's book contains at p. 193, what is I fear the original of Sir James's engraved figure,—and

[1] See Plate 21, p. 254, of *Archæologia*, vol. ii. Also Sir W. Wilde's *Beauties of the Boyne*, and Sir J. Y. Simpson's *Archaic Sculpturings.*

shows that all the mathematical accuracy of the latter drawing, was put into it by the Edinburgh engraver last year.

Questioned for Age.

High intellectuality therefore, we search for at New Grange in vain, in vain.

But then its defenders retreat into the stronghold of its uncountable great age; and its superiority therein, to any remains of buildings known to exist in the more ancient lands of Biblical fame and story.

Of course this exceeding age should be an intensely interesting and noteworthy feature, if it can really be made out. But what are the arguments and facts on which it is supposed to be founded? So far as we can collect them, they are as follows:—

1st.

'New Grange was plundered by the Danes, within thirty or 'forty years of the time that the Great Pyramid of Jeezeh 'was entered by Khaliph Al Mamoun, or in 862 A.D.'

Now what does this prove, except that New Grange was standing when the Danes entered it in 862 A.D.; as they probably did likewise with too many British and Irish houses built either in, or a very short time previous to, 862 A.D.

That moreover, is the earliest contemporary notice that has been yet produced touching New Grange. Whereas of the Great Pyramid of Jeezeh, we have the contemporary foreign notice of Herodotus 1300 years earlier; besides its own speaking and intelligent forms, testifying astronomically for 1600 years earlier still. But there it stops; and does not invade, either by itself, or by helping New Grange,—that most mysterious of Bible epochs, viz. the time before the Deluge.

2*d*.

'The Great Pyramid of Egypt was built in part with iron 'tools, or in the *iron age*; but New Grange was built long 'before that; before even the bronze age; or actually in the '*stone* age of the world.'[1]

This assertion is made up partly of fact and partly of theory. Fact, in that iron tools were known to the builders of the Great Pyramid; and perhaps fact, that at New Grange, what few

[1] This statement may not be intended for the whole of New Grange barrow, but only for some of the carved stones contained in it: though its congeners, the Brittany 'Pyramids,' are very confidently claimed for the 'stone age:' and it is our business here, not to understate any of the claims that either have been, or can be, made for New Grange or any of its fellows, by their admirers.

tools there were, may have been of stone. But theory alone, that a stone, a bronze, and an iron, age successively existed on all the face of the earth; and that each in turn prevailed over the whole of its surface similarly at the same time, and for indefinitely long ages.

The facts then may be allowed: but the theory is the flimsiest ever put forth by mortal man, and is of daily disproof within every one's experience. When Captain Cook sailed forth from Britain in the iron age here, did he not find Pacific island savages still living in the stone age? that is, in the use of stone and bone implements only, and ignorant of metal. Are not some Australians living in that state still? And was one, regular, and uniform, state of civilization *ever* known to spread and prevail equally at the same time over all the nations and tribes and peoples of the earth, from extremest east to remotest west?

Of course not. When Julius Cæsar came to Britain, he found its natives a long way behind, or beneath, the refined and polished *status* of his own nation. And had Alexander the Great visited Ireland, *i.e.*, come from further east to still remoter west, how many grades of civilization would he not have discovered amongst all the several nations and tribes,—barbarians all, in the language of the Greeks,—that lay between.

No reason is there then,—that in the still earlier days of the Egyptian Pyramid-builders,—men in Ireland, if in truth Ireland was then peopled at all, should have been as far advanced in civilization as the dwellers on the banks of the Nile. When builders at Noph, were using iron tools,—bronze was still in vogue in Syria and Asia Minor; Greece was probably in the stone age of Eastern backwoodsmen discoverers; and no Aryan, pre-Celtic, or even Cymbric wave of population was likely to have yet made settlements on Ireland's fertile shore.

3*d*.

'It is asserted that the tattooing style of the ornaments on 'some of the stones at New Grange, and put on divers of them '*before* they were fixed in their present situations,—is of 'that rude, rough, primitive or archaic type, which justifies a 'reference to times, only describable by geological changes, 'dimly shadowing forth to astonished historians whole myriads 'of untold years.'

That some of the said ornamentations are indeed symptomatic of as idle, thoughtless and barbarous natures as man is ever known to have exhibited in all the melancholy chronicles of human savagedom,—is likely enough; but that they are *therefore*, and without further proof, to be considered pre-eminently archaic, and as if man were then raising himself by his own unaided efforts at the beginning of all things, out of a mere bestial and uninspired condition;—is not a necessary conclusion at all. Such an idea is rather, only the wilful prejudice merely of those, who like to deny that a Divine inspiration of man, as related in the second chapter of Genesis, ever took place; and who prefer to believe in spontaneous development, natural selection, the struggle for life, and some other similar animal principles.

In any case almost, of savage-like traces of man's handiwork, executed in unknown ages or circumstances, being alighted upon,—*two* hypotheses seem equally worthy of being tried, to explain them: first, if modern men will, that they were the doing of archaic tribes, the progenitors of subsequently civilized nations; and second, that they were produced by the posterity of once civilized people, but long since ruined by war, or famine, and degraded by unfortunate emigrations, or idolatry and sin.

The first of these hypotheses is the favourite generally with most of our titled authors. Their present country has been on the whole regularly improving in civilization and advancing in wealth, power, and importance, during the last thousand years at least; and they themselves are the latest and most distinguished expression of the fact.

But if some nations are going up, others are coming down, in the world. For where are now the merchant princes of Tyre;—where the strong monarchies, and industrious, warlike, and building, populations of Nineveh and Babylon? Where are the once intellectual inhabitants of Asia Minor; where are vanished the virtues and the genius of the Greeks; and more particularly, what has become of the practical arts, as well as the wisdom, of the ancient Egyptians?

Sir John Lubbock, speaking for the archæologists,[1] declares that man can never lose anything useful he has once discovered; his progress must therefore be continuous, like winding at a ratchet-wheel; that the most advanced state anywhere, is *therefore* the latest,—and the earliest condition of men must have been as low, if not

[1] *Prehistoric Times*, by Sir John Lubbock, p. 475.

lower, than the very lowest state of *existing* savagedom in any land whatever.[1]

Yet were he to make a careful survey of Egyptian monuments from the earliest known times to the present,—it would be found a history of continued degradation in artistic excellence, from a high original status of exact masonry, truthful drawing, majestic sculpture; until now, the descendants of the same men can do little more than build mud huts, and daub over the lintel of the doors some execrable coloured ornaments, or ridiculous imitations of a railway-train of children's trucks, with a black-faced man holding out a red flag. And the melancholy thing is, that these latter-day representatives of the old Egyptians, think these infantine performances of theirs are splendid.

These unhappy men, then, have, without doubt, lost an infinity of things which their fathers knew; and, in a second childhood, are far weaker than their fathers are ever known to have been in their first historical birth.

Are then the idle, scrawling, curving lines, rings, and holes so abundantly engraved on the stones

[1] *Prehistoric Times*, by Sir John Lubbock; concluding chapter.

at New Grange, and other British and Irish megalithic erections,—the work of man in his first, or second, state of childishness?

Asserts Sir J. Y. Simpson,[1]—in a book upon the ring and spiral cuttings in Scotland, Ireland, and France, chiefly:—

> 'From the very earliest historical periods in the architecture of Egypt, Assyria, Greece, etc., down to our own day, circles, single or double, and spirals, have formed, under various modifications, perhaps the most common fundamental types of lapidary decorations.'

But answers thereto William Osburn, specially for Egypt, the first of the countries mentioned, and for whatever formed the examples of architecture there in its *very primal* historic period:—

> 'The earliest monuments in Egypt are, in point of drawing, the best. It is the same in Nineveh and Babylon, as Layard and Rawlinson have repeatedly told me. Whereas, concentric circles, spirals, and other geometrical figures, are the last expiring efforts at lapidary ornamentation among degraded hordes of mankind, just before their lapse into utter barbarism. No such occur on any of the most ancient Egyptian monuments with which I am acquainted.'

But this most mature and well-considered judgment, is confusion indeed to those archæologists,—who would hold that the New Grange, and other similar, stone ornaments were engraved, not

[1] *Archaic Sculpturings*, p. 115.

only long before the Great Pyramid of Egypt had been thought about, but ages also before the creation of Adam. Yet Mr. Osburn firmly implies, from large Eastern experience, that those scrawls are not ancient at all; but are more probably the work of the second and worse childhood of men, long *after* the Egyptian Pyramidal day.

Up therefore once again rises the champion of North European antiquity, to prove the perfect truth of the specially doubted part of his former assertion; or, that the architectural monuments of the *very earliest* historic periods of *Egypt,—are* decorated with 'circles single or double, and spirals:' such things, according to him, forming, 'under various modifications, *perhaps the most* '*common fundamental types of their lapidary* '*decoration.*'[1]

The question is evidently a momentous one for some archæological theories of New Grange. Because, under our searching, but we hope impartial, examination, that monument has lost every vestige of the high intellectuality once claimed for it. Its two first assertions of high, overweening, and ante-Biblical antiquity have already gone by the

[1] Royal Society, Edinburgh, January 6, 1868, and Appendix 3.

board, and now the third and last is hanging only by a thread.

The archæologically-learned Baronet, therefore, had his friends hanging breathlessly on his words, when he propounded, at the Royal Society of Edinburgh, on January 6, 1868—

'that at the Northern Pyramid of Dashoor, as described 'by Howard Vyse and Perring, the stones of which that 'monument is composed, have abundance of concentric circles 'as ornaments; proving the case therefore for times, not the 'very earliest, but still considerably early, of Egyptian archi-'tecture;' and a drawing of concentric circles on one fragment of a stone, was handed round the meeting.

'But his second case,' the eminent speaker considered, 'would be allowed by all the members of the Royal Society,' before whom he read it, 'to be quite unanswerable, and even 'for the *very earliest* Egyptian known example; for it was the 'account given by Professor Piazzi Smyth in the first volume 'of his *Life and Work at the Great Pyramid*, of two concentric 'circles or semicircles on the granite leaf of the antechamber 'to the King's Chamber in the Great Pyramid; such admirable 'concentric figures too, as gave that building over at once to 'the archæologists of rings, and spiral markings.'

And he, Sir J. Y. Simpson, added, that he seized the Great Pyramid immediately, as his own peculiar possession, the moment he saw them.[1]

---

[1] The newspaper report of this address, which occupied its able author nearly half an hour in delivering,—is so exceedingly short, and no other having appeared,—I am obliged to trust to my memory for what I heard, and saw on the occasion; but a very impressive occasion, not soon to be forgotten by any one, and least of all by myself, when beholding my own book perverted to such a use.

The answer to the first case is,—that the stones alluded to, formed no part of the north Pyramid of Dashoor itself, which is a brick Pyramid,—but of the *propylæon* in front of one side of it, and a very different sort of building indeed; probably not earlier than the 15th dynasty, or 1900 B.C., and perhaps vastly later. While as to the concentric circles found on a portion of one of these stones, they are not put in, as are those at New Grange, purely as such and by themselves, but as integral parts of an artistic design, representing a natural object.[1] The design there, is the very frequent one of early Egyptian decoration,—viz., a series of little pillars or posts of a balustrade, formed out of the stems of some reedy water-plants, tied together at intervals of their length, and with the flowering head of the plant plucked off, and stuck with its stalk through part of the vertical bundle, in such a manner as to present the said flowering head, end on to the spectator; and very effectively then designed or indicated

[1] The case in question may be seen as one of eight figs., on a plate without number, but entitled 'North Brick Pyramid of Dashoor,' opposite to p. 63, and described at p. 72, of vol. iii. of Colonel Vyse's *Pyramids of Gizeh*. It was a copy of said figure, out of said plate, which was handed about among the members of the Royal Society, Edinburgh, on the above occasion.

in line drawing, by two or three concentric circles. But these little circles are no more put in for themselves and by themselves alone, than are the two concentric circles of the human eye, in every modern portrait-painter's representation of the 'human face divine.'

And the answer to the second case is,—that a drawing, numerical measures, and description of the supposed discovery of Sir James Y. Simpson, on the granite leaf of the Great Pyramid, are given in the second volume of *Life and Work at the Great Pyramid*, p. 96-100; showing that there are no concentric, or any engraved, circles; but, an anomalous, smooth, semi-oval, raised patch, seven inches wide, seven inches high, one inch thick, terminated rectangularly and rectilinearly below, but bevelled and curved at top and sides. A very peculiar-looking affair no doubt, something mysterious even, and possibly carved *in situ* at a later time;—but totally unable to furnish the whole twenty-one acres of the Great Pyramid's well-dressed original stone surface, with enough of engraved circles, 'single or double, and spirals, 'to form perhaps the most common fundamental 'type of *its* lapidary decoration.'

In fact, one has to wander down, down in history, through the long past Pyramid monumental ages of Egypt, until one arrives at the more recent barbarous and debased Ethiopian structures of Meroe,—before anything like little circles, put in for apparently nothing but idle ornament, can be discovered. Fourteen hundred years of continued degradation, dropped out all the science and most of the art; and at length allowed mere circle ornaments to crop forth on the architecture of the banks of that Nile, whose earlier years had witnessed the glories of the old Empire, in the act, or course, of formation and erection, by the hands of the first Pyramid builders.

And it may be, who knows, that a somewhat similar period of deterioration occurred amongst some tribes of the first *circular* builders; who, flying before their enemies, contending with want and hunger, heat and cold, in their continually western migration,—at last reached the shores of Ireland, some very few centuries before the Christian era;—and both scrawled on, and prepared, pillar stones—of which some were afterwards used by the Tuatha de Danaan, in making up their great burial mound of Brugh,[1] now known as New Grange.

Sculptured scrawls, however, are these, which present no memorials of artistic glory; form no grounds for boasting of the intellect, or wealth, or powerful industry, or warlike renown of Ireland in archæological days; but rather mark the lowest and darkest point to which the race that made them, ever descended. A dismal depth indeed, of misery, ignorance, and woe, which had reached its extremest grade of calamity, when the races which bore it, arrived in that Island of the West, destined of old to be their asylum; and in which, under the blessing of Providence, they have been ever since, slowly but steadily, and we trust will long continue to be surely, emerging. Emerging too, towards far happier days and prosperous times than Ireland has ever known yet, or than it has entered into the minds of the warmest of her patriots to conceive.

[1] 'The nobles of the Tuatha de Danaan were used to bury at Brugh.'—From the Senchas-na-Releo, or History of the Cemeteries, contained in the Leabhar-na-h-Uidhre, a work compiled in the tenth century A.D. *Beauties of the Boyne*, by Sir W. Wilde, pp. 184, 187.

For further Plates of New Grange and its congeners, see Mr. St. John Vincent Day's *Great Pyramid* paper before the Philosophical Society of Glasgow, February 1868.

ARCHÆOLOGICAL DATA—*continued.*

CHAPTER 18.

EGYPTIAN POTSHERDS.

After having fared so badly, *i.e.*, gathered so little, in our search after high intellectual antiquity in the north and west,—it may be prudent to return to some of the Biblical regions of early civilization in the south and the east.

To Egypt then, once more we direct our attention; for, even there, has modern archæological science found something new, and nearly incredible as well as new, to assert.

A few years before his death, the late highly respected Mr. Leonard Horner prevailed both on the Royal Society of London, and the then Viceroy of Egypt, to have certain borings performed in the alluvial soil of the land on either side of the Nile. This was done; and at divers depths of many feet, say 60, or 70, or any other number, burnt bricks and pottery were found. Then, on the assumption that the alluvial soil increases by surface additions of fine mud derived from every annual inundation,—at the rate of, say 3 inches, or 6 inches, or otherwise, per century,—the dates at which these fragments were deposited on the then surface of the ground, is computed to be 16,000, or 28,000, or almost any other large number of years ago.

This conclusion has been attacked by several writers as faulty,—

1*st*, Because irrigation wells being frequent in every field of Egypt; occasionally falling into ruin, and then filled up, while new ones are opened elsewhere,—broken pottery may easily have fallen into them when open, and thus be found now, far below the once historic and contemporary surface of the ground.

2*d*. The Nile having frequently changed its course in the valley, silting up rapidly in flood-time its old channels,—has easily thus placed many things at unnaturally low levels.

3*d*. The burnt bricks frequently brought up in the borings, are considered by Egyptologists to be much oftener Roman, than Egyptian. And,

4th. Some workmen have large proclivities, overseers relate, when not particularly well looked after, for astonishing their highly educated employers, by what they make the machinery under their charge bring up from an unknown depth to the light of day. Whence, almost in anticipation of the care that should have been exerted in these recent Egyptian borings,—Colonel Howard Vyse thus relates in his Pyramid book, vol. i. p. 232, his experiences in Italy,—

'When I was employed in 1832, in an excavation at 'Bosco Tre Case, near Mount Vesuvius, the same piece of 'broken pottery was brought up from a shaft five successive 'times, by way of an inducement to go on with the work.'

To this series of objections may also be added, —that the whole geological principle involved, is wrong in its foundation; for, while it assumes that the pottery or bricks as found now, are still in the very same stratum of soil where they were originally deposited—popular experience or belief tells, that any large lumps of matter, sink through the alluvial soil of Egypt, when well wetted and stirred up by the inundation, as though it were a quicksand.

This peculiar effect seems to arise mainly in

consequence of the extravagantly fine or minute nature of the atoms composing the said alluvial mud ; proverbially black, because so minute in its particles. Small enough indeed are these to float in the air as vapour ; and therefore excessively movable by, and disinclined to settle in, the denser medium of water. But a big fragment of a potsherd, or anything in the shape of a brickbat, of course sinks down through the sludgy medium when it is acting more like a fluid than a portion of solid earth.[1] Hence the depth at which a brick is now found under the surface of the ground in Egypt, is much rather the effect of its sinking in a few years, through an impalpably fine medium slushed with water for several months every year,—rather than because a civilized nation lived on that spot, 20,000 years before the touching story of Joseph had been enacted under the sun.

The Egyptian potsherds, then, of Mr. Leonard Horner, though duly brought up in a recent

[1] See in Dr. Clarke's Travels, his account, in the voyage from Acre to Alexandria, of how the mud of the Nile is occasionally heaped up almost to the surface of the sea by chance combination of currents, forming *apparent* rather than *real*, shoals ; for a ship, he says, striking on one of them, even out of sight of land, is never wrecked, or even stuck fast, so extremely light and mobile is that mud material.

lecture by Mr. Huxley at Sion College, in arguing confessedly and boldly for immense anti-Biblical periods of civilization in Egypt,—are of rather less support, than Egypt's broken reed, of old, to lean upon for any of these purposes.

ARCHÆOLOGICAL DATA—*continued.*

CHAPTER 19.

EGYPTIAN DATE MULTIPLIER.

WRITES the eminent Sir Charles Lyell, in the third edition of his *Antiquity of Man*, p. 383—

'In the temples' (of Thebes) 'are found pictorial representa-'tions of battles and sieges, processions in which trophies are 'carried and prisoners led captive; and if it be true, as 'Sir G. C. Lewis contends, that throughout the historical 'period the Egyptians were a peaceful and never a conquering 'people, the wars to which these monuments would then refer 'must be so ancient as to confer on the Egyptians far higher 'claims to antiquity than those advanced by Bunsen and 'Lepsius.'

An extraordinary paragraph; for, beginning with the Theban temples, which Egyptologists seldom put higher than, if so high as, 1800 B.C.—a facile under-current of argument is indulged in, which presently makes those buildings witness to Egyptian military government, and highly wrought civilization having lasted something more than 10,000 years. Such a rapid mode of getting up an enormous antiquity, which may then be quoted in subsequent discussions against Biblical students with a bewildering effect, is worth while to look into, if only to ascertain how it is accomplished.

The commencement, as already mentioned, is made with the Theban temples; whose dates are moderate, their existence certain, their preservation admirable, and their walls so abundantly covered with sculptures and long inscriptions—called by M. Renan, the Government Gazette of the time,—that they tell whatever tales they do tell (viz., in nine cases out of ten, the contemporary doings of King Rhamses the Great, in 1400 B.C.);—almost as clearly,—as did the London newspapers of their day, the events of the Crimean War.

But then Sir C. Lyell ignores the interpretation of these Theban inscriptions by every Egyptologist; takes the sculpture part only, as representing *a* war, or *some* war,—and prefers to trust for ascertaining *what* war, to a man who hated, despised, and misrepresented the Egyptians; mainly too because, while he knew next to *nothing* of them, he had read up deeply and ardently about the *Greeks*, and blinded himself with their selfish pride and ethnic conceit. This man is no other than the late Sir G. C. Lewis, whom we have already shown to be a sort of anti-Newtonian kind of genius amongst philosophers, in his Egyptian dates. *He* then, Sir G. C. Lewis, simply writes to the effect of the Egyptians having always been, during the historic period, a characteristically peaceful people; and immediately thereon, Sir C. Lyell (quite regardless of the wars both to the north-east and south of Egypt, which national etiquette expected every new Egyptian sovereign to wage),—declares that, *if* Sir G. C. Lewis is right in stating the all-peace characteristic of Egypt throughout its historic days,—the wars shown in the sculptures of the temples at Thebes, must have been *pre-Historic!*

So far, however good,—as to the strict, mechanical inference to be drawn from the approximate *dictum* of Sir G. C. Lewis, quite wrongly assumed to be absolutely true. And if Sir C. Lyell had only been content to take his date for the beginning of such *Egyptian History of Sir G. C. Lewis*, from the same Sir G. C. Lewis himself,—no very serious error in time would have occurred. For, he would then merely have implied, that the wars described on the Theban Temples,—being prehistoric for the Egypt of Sir G. C. Lewis,—must have taken place somewhere before 1500 B.C.; that being nearly the date, wherein that eminent classic scholar places the beginning of Egypt's historical period.

Unfortunately, however, at that most critical point of his argument, Sir C. Lyell suddenly changes his Philo-Hellenic reference Lewis,—for those two thorough-paced Egyptologists, 'Bunsen and Lepsius;' of whom Bunsen claims for the antiquity of Egypt's historic and civilized beginning, something more than 10,000 years B.C. The Egyptian moderate peacefulness then, first seen through Sir G. C. Lewis, being next transferred to, and made to over-ride absolutely, Bunsen's *long* chronology, as it had before done Sir G. C. Lewis's short system,—is, by that one act of the pen, made to begin 8500 years earlier than it would have done, by means of Sir G. C. Lewis alone.

Hoping that this is not a general date-method among geologists,—we now pass on to a distincter literary assertion of some other very long-lived glories of royal Egyptian history.

ARCHÆOLOGICAL DATA—*concluded.*

CHAPTER 20.

EGYPTIAN WRITTEN HISTORY.

IN the course of a paper, read last summer before the British Association at Dundee, its renowned author (Mr. Crawfurd, President of the Ethnological Society of London), adduced, among other arguments for the immense antiquity of man,—both some allusions to the Pyramids of Egypt; and, more particularly, a statement to the effect of the first series of 'dynastic writers' (newspaper report) having lived and flourished on the banks of the Nile, at a date of 8976 B.C.

On inquiring, by letter, his authority for the assertion,—I was frankly referred by Mr. Crawfurd to M. Lesueur's *Chronology of the Kings of Egypt;* and on procuring that handsome quarto book,—published in Paris in 1848, and as a prize essay that had been crowned by the Academy of Inscriptions and Belles-Lettres of the Institute of France,—behold Mr. Crawfurd, in so far, completely justified. For, in the broad pages of M. Lesueur, you may read the names of kings, ruling for thousands of years with that complex and highly organized duplex style of government, of Upper and Lower Egypt; and you may also make yourself acquainted with their hieroglyphic signatures, and the exact lengths of their reigns—even to the minutiæ of odd months and days as well as years,—and this, for all the mighty period beginning with 11,504 years B.C., and ending with Alexander the Great, in the year 332 B.C.

Now, inasmuch as this grand result for proving a long established Egyptian government and history, is brought out with due form and multitude of plates and woodcuts in a first-class prize essay, written to the express requirements of—'*A* '*critical examination of the Egyptian Dynasties,* '*after historical texts* and the national monu- 'ments.'

And as this statement of the theme is further intensified and pointed with purpose, by the addition to it of that most effective phrase of the

French poet, so truly descriptive of mighty and Egyptian-like monuments of stone,—

'Leur masse indestructible a fatigué le temps,'

—seeing all these allusions to monuments,—I did expect that the several main points at least, of that long series of chronological records would be founded on, or referred to, something or other made of, or cut into, enduring stone. Yet after looking the book through from one end to the other, what has not been my utter surprise to find, that there is not a single contemporary monument cited for all the earlier 7000 years of that astonishing series of Egyptian kings.

Nor does M. Lesueur pretend that he has any such monuments. He states perfectly fairly, loyally and truly, that the earliest monuments he can quote are, the Great Pyramid of Jeezeh and its adjacent tombs; while for the 7000 years previous to them, he has nothing but some wretched fragments of MSS.; or scraps of Manetho, and the so-called Turin papyrus,[1] written and in-

[1] The following interesting particulars are from a MS. by William Osburn, author of *The Monumental History of Egypt*.

'As to your question, What was the date of the *original* writing 'of the Turin papyrus?—I answer, as to the existing copy at Turin, 'very late indeed; certainly not before our A.D. 50; probably a 'century later. (A German has found the name of *Christ* in more

dited by not very honest or able scribes some 9000 years after many of the events they presume to describe.

In fact there is no authority, that we can for one moment give credence to (under our settled principles in this book of referring to *contemporary* records),—for all that grand account of Egyptian government and political economy prevailing through a myriad of years; though it was all laid off as fact before the British Association for the Advancement of Science. We may perhaps thank M. Lesueur for his toilsome literary labours in collecting tales that were old to him, though infinitely young compared to the times they ambitioned to speak of;—but, in accordance

'than one of the later chapters.) I said this at Turin twenty years 'ago, when I saw the original. I repeat it after more than fifteen 'years' study of Lepsius's very exact copy. The writers were the 'itinerating Egyptian priests, mentioned by Petronius Arbiter, who 'traversed the length and breadth of the Roman Empire with ass-'loads of images, spells, periapts, and other gewgaws, which they 'sold in the several cities they visited at *remunerative* prices! The 'man who was so befooled as to pay the king's ransom they would 'doubtless demand for 120 feet of papyrus, was one of their con-'verts or perverts, whom they had persuaded to leave the pleasant 'city in Asia Minor, in which he was born, and remove to Thebes, 'in order that he might have his gods about him. Petronius men-'tions this as a case often occurring in his times. The Temple of '*Isis* at Pompeii, the obelisk at Beneventum, and the many pseudo-'Egyptian antiques at Rome, were all the works of these vagabond 'priests.' 1868 A.D.

with the *contemporary* method of inquiry, we must wipe out unsparingly from the proved history of the world, all the names of all his races of kings from 11,504 B.C., down to a few centuries, or years, preceding the Great Pyramid.

A wholesale murder, some may insinuate, but not of the innocents;—for even their best friends make out all those kings to have been, if they ever existed, idolaters through the whole period.

Latest Excavations.

'But does every one agree with you,' many earnest friends may ask, 'in saying that there 'are no monuments in Egypt earlier than the 'Great Pyramid? Was not that building erected 'under kings of the 4th dynasty? Surely then 'there must have been three dynasties previous, 'without going to all those other tribes of previous kings contained in Lesueur?'

The question looks specious. There *ought* to be the monumental remains of three previous dynasties lasting for about as many centuries,[1]

[1] Certain Pyramids near Sakkara are by some authors attached to the 2d and 3d dynasties, but Lepsius in his *Denkmaeler* classes

if Egyptian history really began with the first king of that series of Manetho's dynasties, named Menes; and the archæological world has long been looking forward with extraordinary avidity, to hail the discovery of some contemporary monument of that greatest name in early Mizraite story. M. Renan too says, 'that he has no doubt but that such remains will be found;' and Mariette Bey, the present 'Minister of Monuments' in Egypt, states in his lately published Catalogue of the Museum at Boolak, that his excavators are digging at a certain spot between Jeezeh and Sakkara, in order 'to make to appear the statue of Menes.' But it has not yet come to light.

A startling discovery of another kind has indeed been announced, to the effect, that an engraved tablet has been found on the Pyramid hill,—setting forth how King Shofo, Suphis, or Cheops of the Great Pyramid,—had, amongst

them as of the 5th and 6th; and begins all his magnificent collection of detailed, and chronologically arranged, plates of Egyptian documents, with the quarry-marks of the Great Pyramid.

More strikingly still, in his published *Letters from Egypt*,—he testified that he has found nothing yet, earlier than the Great Pyramid; and that that monument must therefore form the beginning of history, not only for Egypt, but for all the world.

many other public works, 'repaired the figure of the Great Sphinx.'

Such a statement shook the faith of many, as to the long subsequent age of the Sphinx to the Pyramid. For here was monumental testimony that the Sphinx had become dilapidated; and must therefore have been so old, in the days of the builders of the Great Pyramid, that he, King Shofo, had to repair it. But on submitting the inscription to William Osburn,[1] his thorough Egyptological knowledge pronounced it to be, 'nothing but a rigmarole about the god Suphis, 'agreeably with the Suphis and latter-day Osiris 'worship, set up upon the Jeezeh hill in the time 'of the 25th dynasty, or about 600 B.C.'

While as to another statement of two ear-rings having been found, 'which belonged to the wife of Menes,'—Mr. Osburn shows that story to have arisen, partly out of the true discovery of ear-rings with a hieroglyphic on them signifying simply *m n;* but partly also out of the erroneous assumption, that these letters are found only in the name of Mrs. Menes: whereas they occur in

[1] Still living at Leeds, 1868. Author of the *Monumental History of Egypt.*

hundreds of other words; and the ancient Egyptians inscribed every article of their ingenious manufactures, with multitudinous particulars of the events and circumstances of the time,—requiring the use of many words, some of them necessarily including the letters *m n*.

CHAPTER 21.

ARCHÆOLOGICAL RESULTS.

AGAIN then we have to sum up a long list of inquiries,—archæological now, though architectural before;—but the result comes out strikingly closer to what we had previously concluded, from architecture alone.

For here, we are brought back once more to the valley of the Nile, as the scene—all European so-called Pyramids notwithstanding—of the earliest known existing traces of intellectual man. And the earliest monument in these Nile lands, which we can positively lay our hands on,—is the Great Pyramid of Jeezeh.[1]

In so far as that notable fact is concerned, all

[1] 'Nor have I yet found a single cartouche that can be safely 'assigned to a period previous to the fourth dynasty. The builders 'of the Great Pyramid seem to assert their right to form the com'mencement of monumental history, even if it be clear that they 'were not the first builders and monumental writers.

'The Pyramid of Cheops, to which the first link of our whole 'monumental history, is fastened immovably, not only for Egyptian, hail to Mr. Fergusson, on whose architectural history we leaned so much at the commencement of our investigation, and whom we have now found to be so trustworthy a guide both within, and for a considerable space without, the actual limits of his professional science.

He too admires the Great Pyramid in point of construction as much as we can do :—

'No one,' says he, 'can possibly examine the interior of 'the Great Pyramid without being struck with astonishment 'at the wonderful mechanical skill displayed in its construction. 'Nothing more perfect mechanically has ever been erected 'since that time.'

'Nor is our wonder less when we ask ourselves, how it 'happened that such a people became so thoroughly organized 'at that early age, as to be willing to undertake the greatest 'architectural works the world has since seen, in honour of 'one man from among themselves? A king without an army, 'and with no claim, so far as we can see, to such an honour 'beyond the common consent of all, which could hardly have 'been obtained except by the title of long inherited services 'acknowledged by the community at large.'

All this account is as admirable, as it is eloquent, and we must agree with our great historian

'but for universal history.'—Dr. Lepsius's *Letters from Egypt in* 1843.

See also the chronological arrangement of Lepsius's unrivalled collection of folio Egyptian plates, in the *Denkmaeler*.

of architecture perfectly in this,—that the earliest stone building which both he, and we, can find,—is one of the grandest, best-constructed, and most enduring, ever erected by man. We may also agree with him, that the political circumstances under which the arrangements for such an unprecedented building could be entered into, have something very mysterious about them. But when he next considers it to be demonstrated, by the mere fact of so large and excellent a work being constructed then,—that there must *necessarily* have been long and uncounted previous ages of civilized life in the valley of the Nile,—we must take leave to differ; for he is then going entirely on a theory, that architecture must have been *developed*, somewhat according to the Darwinian laws of man's development, by slow continued steps out of the simplest original atom.

To the theory, as such, or rather as a proposition to be inquired into,—there is of course no objection. But we cannot admit its right to pronounce infallible decisions on what took place on the earth in archaic times,—until it has been satisfactorily and scientifically proved, to have prevailed at those periods without a break.

Now if Egyptian civilization did really exist for so long a period, that its duration *before* the building of the Great Pyramid, makes all the years which have elapsed *since* then (though there are more than 4000 of them) sink into utter insignificance,[1]—the question naturally arises, what has become of all the architecture and other contemporary monuments of all those countless ages? And this question is so much the more necessary to ask, seeing that the multitudinous Egyptian monuments *since* the Great Pyramid's day, are generally so admirably preserved by the dry and rainless climate,—that some Egyptologists are inclined to boast, that nothing whether in stone, or even brick, ever decays in that wonderful region,—formed by nature to be, and actually made by an industrious population into, the very monumental land, and enduring record-keeper, of the whole earth.[2]

Mr. Fergusson himself, we believe, admits the difficulty. If his theory is true, there *ought* to have been interminable quantities of pre-Great Pyramid buildings erected in Egypt; and if they

[1] Fergusson's *History of Architecture*, p. 81.

[2] Bunsen's *Egypt's Place in Universal History*.

were so erected, why cannot we find some traces of them? It must surely be a pain, a grief, and an anxiety of no mean order to so honest a man as that able historian, that he cannot show even a single example of the architectural style supposed to be in vogue for ages throughout all Egypt, *before* the Great Pyramid.

He hopes indeed, that that architecture will be found *some* day; and he meanwhile derives a little, though rather sorry, comfort from one feature of his general researches in architectural history, showing—that the *first* specimens of each style, were always and invariably the best of their kind. As, that the Pantheon at Rome, both was the first, and is still the largest, and best, dome in the world; while the earliest of the rock-cut temples of India, are also the most beautiful which that land still possesses.

But the amount of originality in these cases was small; for Theban tombs had been cut in Theban rocks, 2000 years before the Jains began their work in India; and as for the Pantheon, which dates only from the late Emperors of Imperial Rome,—where architects had for ages been brought in from all other countries to contribute toward the glorification of the mistress city of the whole Mediterranean world,—it was no great stretch of invention, from the flat-stone domes and radial arches without number, which both Etruscans and Romans had delighted in for ages, —to rise up to the ideal of a radial dome.

At the Great Pyramid, on the other hand, with its very early date, there was no predecessor building of any kind, pyramid or otherwise, of which we can find any traces—whether in Egypt, or over all the rest of the world. Building in stone, was then a perfectly new art; and the Great Pyramid is far larger and more signal as a lithic structure, than any of the leading examples of long subsequent mere varieties of architecture.

If ever, therefore, there was anything done on the earth, *not* on the development principle,—it was, so far as our present knowledge of all the facts of observation go,—the designing and building of the Great Pyramid. Its sudden manner and perfect appearance on the earth when it did come, can only be paralleled in Pagan literature, with the fabled birth of Minerva, full-grown, learned, and armed at once, on first springing forth from the brain of Jupiter; or, more to the point still, and in the sacred pages of our own Bible,—with what we read there both of the Divine inspiration of man,—by which he became at once a living soul of a high order,—and the communication, at sundry other times, by Divine wisdom, to certain chosen men, of the designs of several buildings according to matchless patterns, measures, and methods of execution.

Opinions on the Final Result.

The bringing forth of such an origin, for the beginning of all archæology and all architecture, has been,—just as might be expected in the present day,—the signal of attack from the 'development' men in natural history, from upholders of the inviolability of nature's laws even by the Framer thereof; from deniers of the Inspiration of Scripture; and repudiators of the truth of any or all the miracles related there.

But what are the arguments or facts brought forward by these gentlemen, against the above result obtaining, as the real and true origin?

One says,

'that we *know* that civilization must always have been a slow 'process; and particularly slow in Egypt, because the Egyp-

'tians being always confined to their narrow valley, had no 'social advantages for seeing and conversing with foreigners.'

Another says,

'that it is so perfectly impossible, according to all theories, 'and even known laws of human progress, for anything very 'great and original to be done suddenly,—that the alleged 'sudden appearance of the Great Pyramid, is an infraction 'of natural laws. In so far, therefore, it cannot be true; or, 'if true, it must be miraculous. But in that case, he, the 'speaker, does not see the object of such a miracle.'

While another still,

'would not limit the definition of a miracle, to his own powers 'of comprehension as to its purpose,—but thinks that any 'interference with the laws of nature is something so momen- 'tous, that all the facts yet ascertained with regard to the 'Great Pyramid should be looked on with suspicion, or at 'least be held in abeyance,—until something more has been 'discovered which may, he thinks, completely alter the case.'

Though we cannot agree with this last objector in fore-casting, as he does, the tendency of any additional further evidence which *may* still come to light, touching the manner in which both civilization's dawn broke upon Egypt and its first great work was executed;—yet we do allow the case to be so extraordinarily momentous, that the utmost degree of caution should be used in weighing and sifting the evidence on either side. More facts may, perhaps, still be discovered in Egypt, and will have their weight; but not an overwhelming weight, because the stock of facts already ascertained with regard to that country is exceedingly large; larger than for any other ancient people whatever; and indeed so large, as to be the frequent point of reference, either for enthusiasm or envy, held forward by the promoters of even Palestine exploration, to a public whom they think not so ready with their subscriptions as they should be.

But meanwhile, seeing that discoveries can neither be made to order on the moment, nor prejudged with success or propriety,—we may do well to put the existing known facts of the Great Pyramid,—as being the first, largest, and best built mere building in Egypt (and perhaps in the world), through a severer scrutiny than it has yet undergone, for its claims, to the still higher attribute of *intellectuality* of design, as well as execution.

INTELLECTUALITY OF THE GREAT PYRAMID.

CHAPTER 22.

ITS SHAPE.

HENCE all our desire now concentrates into this, viz., to ascertain by unprejudiced methods what the Great Pyramid individually is, or was made by labour and art, to be; and also, what it may mean, by being precisely what it is.

This should surely be a pretty safe path, in so far as it can be followed through means of all the most trustworthy measures and examinations actually made at the place, upon the very building itself, in modern and accredited times. And yet certain authoritative objectors appear, recommending us rather to take account only, of whatever vague or indistinct characteristics may be found common to many and various pyramids.

In favour of proceeding thus, it is alleged,

'that the Pyramids of Lower Egypt are more like 'each other, than are European cathedrals like one 'another. But these cathedrals,' they add, 'we 'know, were all built for one and the same mental 'intention. Much more then must all the Pyra-'mids, with their stronger co-resemblances, have 'been built to express one and the same purpose 'of mind. And we shall get at that purpose, 'and test its intellectuality much more speedily 'and powerfully by taking note of all the Pyra-'mids,—rather than by confining ourselves to 'one of them only; whether that one be the 'Great Pyramid or any other.'

A suitable answer to this doctrine, however, appears to be,—

Cathedrals, with their floriated architecture, belong to art.

But Pyramids, when pure, are simply geometrical, and belong to science only.

In art, and under one division of it, say the name of 'ode,' what a crowd of diversities both in composition and sentiment are indulged in there. But in science, what is admitted as a demonstration of—say the 47th problem of Euclid's geometry,—except the demonstration itself; and how totally inapplicable to supply the place of that, would be a general mixture of all the other demonstrations of all the other problems in Euclid, excepting the 47th.

In truth, seeing how intensely mathematical, rather than artistical, the form of the Great Pyramid most assuredly is,—in its original day, simply a huge white Pyramid, with smooth triangular sides;—more exactly and perfectly so too, than all the other known Pyramids;—we might do almost anything else whatever, with more reason, than attempt to explain the Great Pyramid's precisely-built and grandly-unique proportions—by confounding them with the perfectly different proportions of a host of contradictory pyramids, of subsequent ages, different parts of the country, and inferior construction in every way.

We purpose therefore now, to appeal to nothing but hard measured facts at, and in, the Great Pyramid itself; no matter what differently shaped building may stand in some other part of the world. And, if such sturdy data, on being compared with modern exact science,—as now conducted in our best schools,—do not then discover of themselves, and unassisted, some arguments for high intellectuality,—we shall willingly drop the Great Pyramid entirely, and cease to consider it a combined monument of mind and matter.

The Governing Idea.

We have already stated that all the Pyramids of Lower Egypt, when duly restored from present decay to their primitive condition, are quadrangular-based structures ; but differing one from the other in *shape*, *i.e.*, chiefly in so far as regards the proportion of vertical height, to breadth of base.

The particular proportion in this respect observed at the Great Pyramid is found nowhere else. Does it then mean anything particular there ?

Herodotus (445 B.C.) was told by the Egyptian priests, that the proportion was chosen so that the area of one of the triangular sides of the building, might be equal to the square of the vertical height.[1] The proportion, when measured by modern *savants*, is evidently something very

[1] The words of Herodotus on this point are rather imperfect and unintelligible to pure classical scholars ; and were, we believe, first proved to mean as above, by the late John Taylor, author of *The Great Pyramid : Why was it built, and who built it ?* 1859.

close to that; so exceedingly close indeed, that instrumental observation could not have positively disproved the assertion but for, *first*, Colonel Howard Vyse's fortunate discovery of two of the ancient casing-stones of the exterior, *in situ*, and in admirable preservation; and *second*, the measures of the corner angles of the Great Pyramid taken in 1865, from the original corner sockets of the structure.[1]

These and other observations show,—that the angle of the faces, with the base of the Great Pyramid, is not 51° 49′, as required to suit the geometrical idea mentioned by Herodotus; but is nearly 50° 51′ and some seconds. The latter quantity, however, though disproving the first idea (and which cannot be trusted as contemporary information, when received by the Halicarnassian from the lips of priests of a greatly altered religion, and some 1700 years after the building of the monument)—yet brings in another mathematical idea of a still higher character. For the angle of 51° 51′ 14·3″, is the angle which makes the vertical height of the Great Pyramid, bear to the continued length of the four sides of

[1] See *Life and Work at the Great Pyramid*, vol. iii. p. 28.

its base, the proportion of radius to the circumference of a circle.

This quantity when given for the diameter, instead of radius, becomes that most important expression of modern mathematics, known as π;[1] so that the Great Pyramid may then be termed a π-proportioned Pyramid, and stands up in its whole shape a type and lasting memorial, of 'a squaring of the circle' (in the proper way), having been performed ages and ages before the question was ever heard of amongst the schools of philosophy, or the societies of the learned, whether amongst historic Greeks or Egyptians.

The Governing Idea attacked.

No sooner however had this illustrious character been assigned to the observed shape of the Great Pyramid,—than many and many a theorist rushed forward with something else as preferable. When, indeed, the angles which their theories required were several degrees, or even minutes,

[1] Mr. St. John Vincent Day, C.E., has recently called attention, before the Philosophical Society of Glasgow, to another feature of the whole Pyramid, giving the π proportions at once; without the doubling necessary when dealing with the linear proportions only, as above. But as he is expected shortly to publish his essay, I refer readers to his own clear and explanatory pages.

different from the observed, or π-computed, angle, —such ideas could be refused admittance at once. But when in the course of last autumn, a thought was brought up requiring within a few seconds only, the very same angle as the π hypothesis,—a new difficulty had arisen, for the case could not be discriminated then by observed angle alone.

That this new view,[1] was promulgated by Colonel Sir Henry James, R.E., showed that so able a judge as the Superintendent of the Ordnance Survey, considered the observed angle of the Great Pyramid, to be by this time well measured, and very close to the truth,—though he differed as to the governing idea in the mind of the original designer.

That idea he considered to be,—merely taking for the vertical height of the Pyramid, $\frac{9}{10}$ths of the semi-diagonal of the base. And, as that gives for the height, a quantity different by only $\frac{1}{3000}$th of the whole, or two inches nearly on the entire height of the Pyramid, from what is given by the

[1] First discovered by Mr. Petrie, C.E., and communicated to me in MS., amongst six other sheets of Pyramid details, October 16, 1867; but independently discovered very shortly afterwards by Sir Henry James, and by him, both sent to me and afterwards published in the *Athenæum*, 16th November 1867. See Appendix 1.

π hypothesis combined with the usually received absolute size (which latter too is not certainly known within a whole handful of inches),—why, two equal competitors seemed here brought up, for the honour of having been the mental intention of the primeval architect.

The occasion was one for cool criticism and patient investigation; for the coincidence of either idea, or perhaps of both, with the observed angle of the Great Pyramid,—might be an affair of accident only. Sir Henry James however, had no such difficulty; for he at once gave the decision absolutely in favour of his own hypothesis, and announced that 'he *had* discovered the mode of 'construction of the Great Pyramid.'

The grounds of this decision seemed to be, that the $\frac{9}{10}$th semi-diagonal, was a *simpler* idea than the π proportion; and some modern gentlemen have a peculiar pleasure in trying to show, that the designer of the Great Pyramid was of poor mathematical powers, and mean constructive attainments; or, as unable to conceive a high thought, as to put it into practical execution.

Now there is no harm, but, on the contrary, much positive merit, in always trying *first*, the simplest possible hypothesis, for explaining any given phenomenon. But we have no right to favour such simple hypotheses, by keeping out of view any additional observations of collateral facts. Such facts did exist in this case, in the shape of the angles of the very curious system of descending and ascending passages inside the Great Pyramid,—for they had already been shown to be conformable to a further development of, or to constitute a corollary to, the π theory of the outside.[1]

This by itself, was more than doubling the claim to *intention*, for the π hypothesis; though the measurement of the precise angle is a difficult one. Yet at the instant of despairing what to search for next, or that could be better—behold Mr. William Petrie, alights upon a discovery which seems to make the matter plain for all men now to read easily, even as they run.

Testimony of the Trenches.

Huge, deep, and long rectangular cuttings in the rock of the hill, lie before the eastern face of

[1] *Life and Work at the Great Pyramid*, by C. Piazzi Smyth, vol. iii. pp. 34-44, and 219.

the Great Pyramid; variously destined according to previous authors, but denominated by myself 'azimuth trenches;' because, while two of them lying north and south, nearly (or within 19′) mark a meridian line,—the others radiate out from the middle point between these two, at horizontal angles representing the summit and foot of the Great Pyramid.[1] And for several reasons mentioned in *Life and Work*, I ventured to consider these trenches to have been originally incised, for instructing the masons in the exact angular character of the very mathematically formed building they were engaged on, and *while the work was in progress.*

This hypothesis seemed to explain the angular differences of the trenches well, but said nothing about their position with regard to the Pyramid, or their length. Now, however, these matters are answered by Mr. Petrie; who says, that if you take the Great Pyramid, as it was when in masonry progress, or without its final casing-

[1] At p. 126, vol. ii. of *Life and Work at the Great Pyramid*, are given measures of azimuth trenches, by which these points may be tested: while the further information is now tendered, that the sides of the Pyramid referred to there, are the present broken sides, *within* the area of the casing-stone socket marked sides.

stone film,—and if from the centre of the then base, you draw its proportionate π circle;—the conjoined axes of north and south azimuth trenches, will form a tangent to that circle, at its most protuberant point in front of the middle of the east side.

And further, if from the points towards the north and south extremities of the east side of the base, where the π circle cuts into the area of the base, you draw rectangular offsets from that side, eastwards,—these offsets will be found to define the places of the admirably square-cut outer ends of both north and south azimuth trenches, with as much accuracy as the present standing and broken sides of the Pyramid admit of, in their measurement.[1]

No less than three special points of the π theory, being thus found marked, and in so magnificent and easily legible a manner, to all men—whether using microscopic instruments, or ordinary plain vision and common sense,—we may hope that that theory is now firmly established as having really been in the mind of the designer.

[1] *Life and Work at the Great Pyramid*, vol. ii. p. 126.

N

INTELLECTUALITY OF THE GREAT PYRAMID—*continued.*

CHAPTER 23.

ITS NUMBERS A COROLLARY TO SHAPE.

THE previously mentioned satisfactory conclusion as to *shape intention*, was no sooner gained in our investigations,—than various objectors appeared against the ancient architect of the Great Pyramid, on this very account;—one of them asserting with no little emphasis,—

'What a much better figure a *cone* would have been, than 'any Pyramid, to express the π relation of diameter to circumference of circle!'

While another gentleman criticises severely both cone and pyramid, on important 'artistic grounds.'

'The upper part of these two classes of figures,' says he, 'retreat too rapidly from the eye, and are thus never seen of 'their true size or with their full dignity. Whereas, had the 'architect only been properly advised to adopt a cylinder in 'place of a cone, or a *rectangular* mass in place of the *acute-* 'angled pyramid,—the superior vigour of outline of these 'forms when realized on so grand a scale,—would have pro- 'duced an effect positively sublime.'

At present we need only remark in answer, that although the π hypothesis may have been proved to have been in the mind of the designer when planning the Great Pyramid structure,—we have arrived at no proof, that that one thought was the *only* thought there. And although cones, cylinders, and cubes of certain proportions may be made to designate the π ratio,—yet they do it with a decided difference of accompanying qualities. A cone, for instance (taking the base, as one side), has only *two* sides or surfaces; a cylinder, three; and a cube six; but a quadrangular-based pyramid, reckoned after the same manner, has five.

Now if,—besides the π proportion of the Great Pyramid, and an acknowledged geometrical figure, —that building was intended to contain some system of numeration, or of symbolization by number, wherein five was to be, if not the very *radix* itself, at least a number of frequent reference and employment,—it is plain that a square-based π pyramid has incomparable advantages over a π cone, cylinder, or cube; in fact, it is the right figure in the right place.

The peculiar and characteristic fiveness of its sides or surfaces, moreover, repeats itself in the similar fiveness of such a pyramid's angles or rather corners (viz., four below and one above). So that then, from such a figure, with its five sides and five corners, the collective number of ten, is thence derivable: even very much as all decimal arithmetic is usually held to be founded on the two, so-called, five-fingered hands of man; though each has really only four fingers and one thumb.[1]

The Facts derided.

That any importance could be attached to these two groups of *five* in the Great Pyramid,—was recently the special subject of attack, denouncement, and ridicule before the Royal Society of Edinburgh, by a gentleman,—who quoted as one of his most convincing instances of the absurdity of what he denounced,—the hint given to him,

[1] Somewhat similarly of the pyramid's 'five sides,' four are triangular, and one is the square base; while of its five corners, it has four similar ones below on the foundation level, and one above, in a very elevated and distinguished position.

he said by a friend, to the effect,—that the very room in which the Society was then meeting had five windows, and five divisions in its ceiling, repeated twice, besides ten chairs, sometimes five benches on either side, and certain other less essential detail.

So it had; but were they as dominant features of the whole building they belonged to, as are the two pervading fives at the Great Pyramid?

At the Great Pyramid, the whole surface and the whole substance of the building (and that the largest and most massive as well as most ancient in the world), combines in making up these two geometric sets of five, and apparently nothing else extraneous besides. But at the Royal Institution building, wherein the Royal Society of Edinburgh meets—what do we find? An ornamental hollow structure of the present century, erected on a rubbish-heap; and serving a multitudinous set of purposes for many men, and different societies. Outside, you see porticoes, fluted columns, cornices, pediments, lions' heads, chimney-pots, windows many, sphinxes mounted up on the frail roof, Queen Victoria's majestic figure in stone; volutes of Grecian vegetation, protrusions here and recesses there,—nothing certainly which leads an observer to consider, that a remarkable, or indeed any, purpose of five-ness runs through all the building and forms its most distinguished, most pervading, and most easily recognised type, uninterfered with by any other number.

And inside,—we find a side suite, only, of rooms for the Royal Society, a central one for the Antiquarian Museum of Scotland, and others upstairs, sadly dominating those of the Royal Society, for the School of Design under the Board of Trustees of Manufactures; with divers little chambers for officials; and with almost every diversity of windows and ceilings, chairs and benches amongst them.

Really therefore, had not the case been brought forward before a very full meeting of the most scientific society of Scotland, and not only treated with marked favour from the Chair, but held by many rather thoughtless persons present, to have proved, as they termed it, '*a smasher*' to the importance of the Great Pyramid,—I should not have thought it worth notice. And even now can only compare it to some innocent youth picking the fives out of a multiplication-table, containing numbers up to one hundred and forty-four, and declaring that the whole table was formed for the fives alone; or at all events, that that little bit of modern paper, containing some fives, amongst other numbers up to 12 times 12, possessed all the excellences, in the way of grand, primeval fiveness, claimed for the Great Pyramid.

When any earnest inquirer shall have discovered, not only a mere page of various figures, of which two or three are fives,—but a large folio volume wherein every page throughout, except perhaps on one small portion, shall be filled with fives, and nothing but fives,—he will have discovered both something very unique in numerical literature,—and something that may give him an appreciative idea of the enormous amount of fiveness extending uninterrupted and dominant through all the Great Pyramid's mighty bulk.

But until that volume is found, or until another building shall be discovered somewhere in the world, as pure, simple, and extensive, in its fiveness, and making as much use of its fiveness afterwards in numerical references towards the noble things of the earth, and the high things of heaven, as the Great Pyramid will be presently proved to do;—we suspect that that structure's remarkable design, in the world's early ages, has not been very much injured, or altered, by this late attempt to hold it up to modern ridicule.

Subsidiary Numbers.

Though in a lesser degree, than fives,—four may also be considered as a Pyramid marked-out number; because four is the number of the sides of the square base, and four the number of the triangular faces or flanks.

Three and seven have likewise been suggested by William Petrie, as numbers connected with the Great Pyramid. Though, too, at present these numbers are only to be dimly seen, in a rather abstruse manner (notwithstanding that they are general and universal to the whole surface), yet they have the discriminating merit of belonging to the Great Pyramid only, of all the Pyramids of Egypt, and the world.

This useful property comes to the Great Pyramid,—without altering the previous fives or fours in the smallest degree,—through the agency of the peculiar π figure; for when that π fraction is duly understood,—and it has now been computed to the extent of 609 decimal places of our denary arithmetic,—it is found, that while all other digits occur ordinarily, 3 and 7 occur most peculiarly and exceptionally, yet with some apparent law determining them to do so.[1]

The numbers 3 and 7, then, cannot be seen instantly on the surface of the Great Pyramid; though they may be understood to be there, most securely, through the latest teachings of modern science regarding the really expressed quantity π. This sort of presence indeed, of 3 and 7, if taken by itself, would be, perhaps, of no very great mental importance; but just as with the equally necessary fives, tens, and fours of the visible surface; so with regard to the threes and sevens of its essential figure,—it is the *use* which the Great Pyramid system subsequently makes of them for high and noble ends, which forms the chief claim for mental intention on the part of the designer, when he adopted the π-proportioned, quadrangular-based Pyramid; and some few examples of that use we shall now endeavour shortly to set forth.

[1] See postscript, p. 602, vol. iii., of *Life and Work at the Great Pyramid*. Also a fuller recent paper by W. Petrie.

INTELLECTUALITY OF THE GREAT PYRAMID—*continued.*

CHAPTER 24.

ITS SIZE.

THE *shape* of an accurately formed building to be, such as the Great Pyramid 4200 years ago, having once been settled, with the attending numerical characteristics,—its *size* would evidently next occupy the attention and require some new mental determination on the part of the designer. But seeing that we have not him himself before us to question on this point—let us examine his work in order to ascertain, as a first step, how large he made it.

As already abundantly described elsewhere,[1]—the only present existing traces,—by which the full extent of the Great Pyramid in its early

[1] See *Life and Work at the Great Pyramid*, by C. Piazzi Smyth, vol. iii. p. 124.

days, seems to be satisfactorily marked out,—are, *first*, the sockets worked into the flat foundation rock, and intended for the corner-stones of the outer casing; showing therefore the length of the sides of the original base,—and *second*, in connexion therewith, the angle of rise of the triangular faces, from the sides of that base. For with the distance known from socket to socket, and the angle of the faces,—of course either the vertical height, length of corner-lines, or any other feature, can be immediately computed.

The angle of the faces we do know, probably, with sufficient closeness from many measures, to justify us in assuming the theoretical π angle as the one intended; and that can be computed to all required refinement. But the true horizontal distance from the outer corner of one socket, to the similar outer corner of another,—or rather, the mean of all four sides similarly measured,—is still, for accurate purposes, an unsolved problem. And what is worse, it will always remain so, until some strenuous efforts are made—by the rich and powerful of the world—to clear the intervening ground of the hills of rubbish which now prevent scientific men from obtaining the desired accuracy.

Of all observers who have yet attempted to get this measure of socket to socket, from the French Academicians in 1799 to Messrs. Aiton and Inglis in 1865,—their results have been somewhere between 9168 and 9102 British inches.[1]

Somewhere between these two lengths therefore, the true one may be expected to be found; the large difference between them, being not due to any real alteration in the size of the compact rocky hill of foundation,—as rather strangely inferred by a medical gentleman before a learned audience recently,—but merely to the errors of the observers themselves; and which errors are almost as likely to be positive as negative. Exactly *where* the true length is likely to be found, between these extremes,—is a nice question of mathematical probability, involving so many unknown quantities, that it is not likely to be settled speedily. In the work already mentioned, I have followed, for reasons there detailed, a method which gives 9142 British inches as the probable truth; and shall therefore use it now, though with the due *proviso* that it cannot be considered certain within limits of perhaps ± 25

[1] *Life and Work at the Great Pyramid*, vol. iii. p. 125.

inches.[1] These limits, however, only amounting to the proportional quantity, on so large a building, of $\frac{1}{360}$th of the whole length, and exceeding the next largest Egyptian Pyramid by 500 inches, or $\frac{1}{18}$th,—may enable us to approximate to something important.

Depreciating Theories.

From time to time ideas have been published as to why the Great Pyramid was made of the precise size we find it, sometimes giving rather learned grounds; and sometimes declaring the cause to have been, merely the result of ignorant, overbearing pride of a tyrannous king, striving after brute magnitude, without regard to quality, mind, or soul.

[1] William Petrie, after a much more extensive, critical, literary, and scientific investigation of the authorities, has concluded 9165 to be more probably the truth; and he may be right; I would even say that he has a high degree of probability on his side. But nothing short of removing the rubbish obstructions at the Pyramid, and taking new measures with improved apparatus, will be sufficient to settle the point; and there is no reason why the length of the base sides of the Great Pyramid should not be measured with all the microscopic accuracy of Colonel Colby's 'Compensation bars;' which have already been employed in several countries to measure the length of base-lines with extreme success,—but at vastly more *expense*, both of time, labour, and money, than has ever yet been bestowed on the base-mensuration of the Great Pyramid; and without which, no first-rate result in accuracy can ever be expected.

In the midst of these uncertainties, a few Egyptologists have been so hardy as to declare, that there was *no reason at all;* and that the Great Pyramid, in a manner, simply *grew* to its present size, and stopped there. For, according to the building theory framed for subsequent pure sepulchral pyramids, by Mr. J. Wild, architect, and Dr. R. Lepsius,—each ancient Egyptian king, on his accession, began the pyramid which was eventually to be his tomb, in the following manner :—In the first year of his reign he excavated a subterranean burial-chamber, with an inclined entrance-passage; and placed a few blocks of squared stone over it. In the second year he added another layer of stones vertically, and increased the lower one proportionally sideways;[1] and so went on, every year, until he died. Then his friends entombed him in the subterranean burial-chamber; and finished his pyramid outside, by casing it smoothly with bevelled stones.

Hence the size of any Pyramid, according to this theory, is merely a consequence of the length of reign enjoyed by its king.

[1] See Plate 10 in *Our Inheritance in the Great Pyramid,* by C. Piazzi Smyth; A. Strahan and Co., 1864.

Whether this was the case with subsequent pyramids, we do not pretend to say or dispute,—but altogether doubt the application of such a principle to the Great Pyramid, for the following amongst other reasons:—

1*st*. On this method, the subterranean chamber—the first begun,—should be the completest part of the whole Great Pyramid. But in place of that, it is absurdly unfinished, and only half cut out of the rock.[1]

2*d*. Herodotus gathered, upwards of 2300 years ago, a most decided and extensive tradition in Egypt,—and in precisely a matter where popular tradition may be trusted,—to the effect that the building of the Great Pyramid was an unprecedentedly great and well-organized work from the first: no less than 100,000 men being employed upon it at once, regularly relieved every three months by others. These enormous gangs too, were kept very energetically to their tasks, while the building was in progress; though that was only for a limited number of the many years during which the king reigned. And,

3*d*. There is no appearance of changing times,

[1] See Plate 4, vol. iii. of *Life and Work at the Great Pyramid.*

or varying fashions, no sign of year-to-year work long drawn out, no small beginnings afterwards expanded or patched up with other things, throughout all the masonry of the Great Pyramid, so far as it has yet been examined. But everything on the contrary indicates that its designer knew beforehand *how* large he was going to make the monument, and prepared accordingly. In fact, the Great Pyramid,—in so many other points more truly and purely a pyramid, than any other of the pyramids of Egypt,—was also pre-eminent in *Promethean*, rather than *Epimethean*, symptoms and accompaniments.[1]

A Determined Attack.

An idea not quite so slighting to the scientific claims of the Great Pyramid and its architect, as the previous accidental or spontaneous hypothesis,—was that published by Colonel Sir Henry James, last November. For that eminent officer did in so far admit the unity of the building, and the Prometheanism of the design,—but could find no better reason or higher motive for such immense organizations of labour having been set

[1] See p. 87.

into train, in an early age of the world, than merely, this;—viz., that the Great Pyramid might have one of the sides of its square base, equal in length to 360 cubits; each of them being of such a size, as now to measure 25·488 British inches long.[1]

The length of base-side, however, hence resulting, is not, as stated by Sir H. James (in its equivalent of feet), 9168, but so much as 9175·68 British inches; and this quantity lying outside all the measured lengths just detailed,—cannot be accepted without demur, or at least not until it receives some further confirmation. We ask, therefore, what was the importance to the builder or designer of the Pyramid in using the number 360; and what are the merits of a cubit of the exact length stated, or 25·488 British inches?

As to the number 360, Colonel Sir H. James does not give any testimony in its favour, but simply employs it. We, therefore, finding it neither a very round number, nor marked out anywhere very clearly by nature, nor deducible from the Great Pyramid numbers already laid down,—cannot receive it as necessarily dominant; and we rather suspect that its first use for the

[1] *Athenæum*, London, November 16, 1867, p. 650.

degrees of a circle, dates only from Babylonian times, some 1500 years after the building of the Great Pyramid; and when *sexagesimal* division, absolutely unknown to the Great Pyramid, was coming into favour.

And, as to a cubit of 25·488 inches long,—seeing that nothing is stated by Sir H. James in favour of that being an important natural, or scientific, standard,—its only chance of recognition depends on reason being shown, for such a length of linear measure having been employed, and very highly esteemed, in ancient Egypt, at the date of the Great Pyramid's foundation.

This essential point Sir H. James does attempt to show; but we fear decidedly fails in, because,—

1*st*. His own measure of an *ancient* Egyptian cubit, unites with the multitudinous determinations long since printed for the same, by Sir Isaac Newton, M. Jomard, Sir Gardner Wilkinson, and almost all Egyptologists; *i.e.* making it something close to, perhaps within a few hundredths of an inch—his own is within a thousandth of an inch—of 20·7 British inches; an absolutely different quantity from 25·488 British inches.

2*d*. He finds an Arab cubit recorded about

two centuries ago, as being 21·888 British inches in length; and on adding $\frac{1}{8}$th part thereto, he finds it come out 25·536; or moderately close to 25·488 inches. This is true. But the plan, far-fetched and more than doubtful in propriety even as a method of using the said Arab cubit in its own day,—and which cubit some modern authors quote at a much less figure, or as 21·39 to 21·32 inches only,[1]—is utterly powerless to bridge over that immense gulf of time which separates 1600 A.D. from 2170 B.C.; or to prove, that the builders of the Great Pyramid necessarily employed a cubit of 25·488 inches, in laying out the base of their monument, *because*, forsooth, certain Arabs 3700 years afterwards, used a cubit between 21·888 and 21·32 British inches long! And,

3*d*. Sir H. James falls back on 25·488 inches, being the length of the *modern* 'derah' or land cubit of Egypt. His authority for the statement is a little Manual on 'the weights and measures of all nations,' by Mr. Woolhouse, published in Weale's series of rudimentary treatises. The

[1] This cubit is that of the Nilometer at Rhoda, stated by Sir Gardner Wilkinson, in Murray's *Egyptian Handbook*, to be 21$\frac{7}{18}$ English inches long; but quoted from M. Coste at 541·5 millemetres.

excellent mathematical author of that Manual, was never, we believe, in Egypt; was unable, when applied to recently, to give his references for the numbers: and stands directly contradicted, by what is published in the works of two men so admirably conversant with modern Egypt, as Mr. Lane and Sir Gardner Wilkinson.[1]

[1] Mr. Lane writes, at p. 571 of his *Modern Egyptians*, 5th edit.:—

'Of the measures and weights in Egypt, I am not able to give an 'exact account; for, after diligent search, I have not succeeded 'in finding any two specimens of the same denomination perfectly 'agreeing with each other, and generally the difference has been 'considerable.'

'The *diraa beledee* (or cubit of the country—the common 'Egyptian cubit), which is used for measuring the linen, etc., manu- 'factured in Egypt, is equal to 22 inches and two-thirds.

'The *diraa Hindazeh*, chiefly used for measuring Indian goods, is 'about 25 inches.

'The *diraa Istamboolee* (or cubit of Constantinople), which is used 'for measuring European cloth, etc., is about 26 inches and a half.

'The *Feddan*, the most common measure of land, was, a few 'years ago, equal to about an English acre and one tenth. It is 'now less than an acre.'

Sir Gardner Wilkinson, in Murray's *Egyptian Handbook* for 1867, writes—

'MEASURES OF LENGTH.

'1 Drah beledee, or cubit, equal to 22 to 22$\frac{3}{4}$ inches English.

'1 Drah Stamboolee, . equal to 26 to 26$\frac{1}{2}$ inches English.

'1 Drah Hindazee (for cloth, etc.), equal to about 25 inches English.

'LAND MEASURES.

'22, formerly 24, Kharoobeh or Kubdeh, . .	=	1 Kassobeh, equal to from 11 feet 4$\frac{1}{2}$ inches to 11 feet 7$\frac{1}{4}$ inches English.
'13$\frac{7}{8}$ Kassobeh or rods,	=	1 Keerat.
'24 Keerat, or 333 Kassobeh,	=	1 Feddan or Acre.'

These authors indeed agree well together, in setting forth the present 'cubit of Egypt,' to be from 22 to 22·67 inches long. And though Indian and Turkish cubits of 25 and 26·5 inches respectively, are used in the shops for Indian and European silks and cloths,—the only other cubit that can now be considered national in Egypt, or as connected with the 'land,' and likely to be tolerably permanent,—is that of the Cairo and Rhoda Nilometer; and is stated to be from 21·39 to 21·32 British inches long. In fact, approximating much more to the ancient cubit of 20·7 inches, than to this mystical one of 25·488 inches; which latter, too, is not proved to prevail now, or to have prevailed, as the country cubit and 'land measure cubit,' of Egypt, at any time whatever,—either ancient or modern.

Hence we are compelled to try some other hypothesis, than Colonel Sir H. James's, for having decided the Great Pyramid's absolute bulk.

Elevating Ideas.

The method of depreciating the mind and aims of the designer of the Great Pyramid, having failed in producing a sufficient, or even a probable, reason for the choice of its particular size;—let us next try, though only as a tentative proposition, something more respectful to that unknown, yet evidently excelling, personage of primeval times.

In this point of view we would remark,—that the method of looking to one side of the base, when there are three other sides to dispute with it,—is not the neatest principle on which to describe the size of a Pyramid (the angle of whose faces has been already determined). We should rather look to secure a feature which is unique, central and governing to the whole,—and such a feature is,—the vertical axis.

This vertical axis moreover commends itself to us, at the Great Pyramid, by its being the π radius of the pervading symbolization of its shape. Nothing therefore is more suitable, than that this π radius should have been made of such a length or on such a scale of size, as to testify something else scientific. Let us ask therefore with what grand, or small, radius in nature, does the π radius of the Great Pyramid measure itself accurately, and in even terms of Great Pyramid numbers?

William Petrie, under guidance of the general bearing of all the system of symbols which he and others had found in the Great Pyramid, was led towards the close of last year to *expect* that such commensurable natural radius should be, the mean distance of the earth from the sun; and that the numbers 10 and 9 should be pre-eminently included; though it was some time before he could fully believe, that the mean solar distance from the earth was really, simply, and precisely 10^9, (*i.e.* 1,000,000,000) times the vertical height of the Great Pyramid.

In a general manner we may remark, that this idea fulfils a vast amount of scraps of Eastern tradition, connecting the Pyramid with the sun, or with solar light. And while there was an ancient fancy of mythical importance among the Greeks and Romans (but descended to them from earlier times),—as to the greater glory of the Pyramid when the sun rose to so notable a height in the sky at noon with advancing summer, that then, as they expressed it, the Pyramid 'devoured its own shadow,'—or, according to a phrase since found in use among the Peruvian worshippers of the sun,—'when the sun sets with all his beams on every face of the Pyramid at once,'—let us ask how the sun does shine on the Pyramid under those circumstances?

Why, it shines simultaneously on all the five corners, and four triangular faces,—or nine chief parts; which fact, combined with the abiding decimal system of the whole building, makes 10^9 a number worth investigating. All the more too, when the Pyramid is of such shape, that, with an error of only $\frac{1}{3000}$th, it rises nine parts towards the sun, for every ten parts that it stretches along horizontally, in the direction of its base's semi-diagonal length.

Now the vertical height or π radius of the Great Pyramid, has already been shown, in *Life and Work*, to be about 5819 (subject to error of determination of perhaps ± 16) British inches; and that quantity being reduced to miles,—the term in which sun's distance is usually expressed,—becomes ·09184. This again being multiplied by the 10^9 of the above formula gives 91,840,000 miles ± 260,000 miles, for the Great Pyramid's expression of the sun's mean distance from the earth.

How then does this agree with modern science?

Were we still living in the state of knowledge prevailing from the beginning of the present century, to about eight years ago,—the answer of every astronomical treatise throughout Europe and America, would have given, for the true determination of the sun's distance, as computed from the last observed transit of Venus, the exactly defined quantity of 95,293,055 miles.

In such case the Great Pyramid quantity of 91,840,000 ± 260,000 British miles, would have been condemned as utterly too small. Yet the Pyramid would not have been wrong; its symbolization would only have been discovered too soon, and before the world was prepared to receive or able to appreciate it.

Of all the material elements for supporting life on the earth, none are so dominant as those depending on the light and heat of the sun; and of all problems in natural science, none is so important as the true distance of the sun. Upon that depend our knowledge of its weight and size; the weight, sizes, and distances of the planets; and even the masses, distances, and light-power of the fixed stars themselves. The sun's distance therefore is *the* great problem; and modern science has spared neither time nor expense in endeavouring to settle it. But the distance is so great, and our base-line for measuring it (the earth's breadth) so comparatively small,—that modern science has found her telescopes, micrometers, divided circles, and mathematical methods, all taxed beyond their powers in accurately determing *what* the distance is.

Hence on new observations being made in 1862 (at the opposition of Mars) the solar distance was found to be, by those results, not 95,293,055, but more nearly 91,430,000 miles. While again, extensive recomputations of the old observations having been made during the last year, in America, by Professor Simon Newcombe,—especially by employing correcter modern-determined longitudes of the places of observation,—the most probable quantity is now stated to be 92,380,000 miles.[1]

Modern determinations then of the mean solar distance, though pushed on by the whole strength of modern science, are only oscillating on either side, and at every fresh attempt are approaching

[1] Mr. Petrie's result, from his investigated height of the Great Pyramid, and without knowledge of Professor Newcombe's numbers, amounts to 92,093,000 miles.

closer and closer to the Great Pyramid symbolization in days of old; and not until the next *good* observing transit of Venus over the sun's disk, or in 1882 A.D.,—will any result be probably attained, that can be confidently relied on, as more accurate.

General Considerations thereon.

The state of things just mentioned, is worthy of the gravest attention. Modern science, on account of the excessive difficulties surrounding the case, cannot determine the important problem of the sun's distance within moderately close limits. It is only within the last seven years, that its numbers have been even respectable. A hundred years ago, the best approved scientific result was no less than 10 millions of miles in error. Two hundred years ago, it was 66 millions of miles in error; and for the 1900 previous years, it was 87 millions of miles in error; and previous to that again, 92 millions of miles in error; or more than 99-hundredths of the whole quantity concerned.[1]

[1] These few particulars are drawn from an extensive paper on the subject, recently communicated to the Royal Society, Edinburgh, by Mr. William Petrie, C.E.

That was at the dawn of human science, the origin of the Grecian schools, and the beginning of man's thoughts being turned towards ascertaining for himself any of the data of the heavens in number, weight, and measure. And yet, 1700 years earlier still, or in 2170 B.C., a building was erected on the surface of the earth, containing in its chief substance a permanent expression of that quantity,—put in without hesitation, or delay, or apparently any sensible error!

Some objectors will indeed say,—that it is absolutely impossible for men who lived 4000 years ago, without telescopes, and without observatories, in the Southern hemisphere as well as the Northern, to have approximated in any degree to the real distance of the sun. And in so far we allow they have reason on their side; for no mere men, by science of human cultivation, —which *is* an affair of gradual progressive development,—ever could have done anything of the kind accurately; and they never did so, at any date of the world's history earlier than the last century. The ancient Egyptians too, themselves, knew nothing of the sun's distance, and hieroglyphed the earth as a flat.

But then, while the Great Pyramid facts are certainly such as we have described;—the Bible tells us, that in very early historic days, wisdom, and metrical instructions for buildings, were occasionally imparted perfect and complete, for some special and unknown purpose, to chosen men, by the Author of all wisdom. For those therefore who believe the Bible, there does seem a possible way out of the difficulty,—though to those who do not believe in that Book's inspiration words,—and trust only to Darwinian and Crawfurdian theories of human development by innate powers out of an atom, or an ape, or a bestial savage,—the Great Pyramid, a silent but majestic spectator of the whole historic period of man as yet,—is constantly giving forth a witness which they cannot explain.

We do not however at present wish to enter on the argument, of whence came the knowledge which designed the Great Pyramid to be what it is, as ascertained by recent measure;—because we are still only at the beginning of the series of testimonies which that building has to give. Enough therefore to add now, that while modern science is still going on improving its determinations of the sun-distance,—and while there may be obtained, by renewed measurement at the Great Pyramid, some slightly different value for its vertical height,[1]—the formula admits of *no* alteration. 10^8, for instance, would not suit the circumstances of the Pyramid when the sun is dominant, and would make the solar-distance so small as to thrust the earth into the fervid heat inside the orbit of Mercury; while 10^{10}, equally unsuitable in principle, would expel our planet into regions of cold beyond the orbit of Saturn.

10^9 is the only simple expression, which seems to be at once suited by nature and art to be the expression, and at the same time to give the true distance of our earth from the sun. That expression, we mean, when combined with the height bestowed on the Great Pyramid by its builder. And such height, not only realizes the solar problem under the appropriate formula, but is a very unique and remarkable height in itself.

It is the greatest height, for example, of any building ever yet erected by man. For, though the

[1] While my own idea of the Great Pyramid's height, has already been given as 5819 ± 16 British inches; William Petrie inclines to a mean quantity of 5835 of the same inches, with a much smaller probable error.

first known specimen of all human architecture, yet the Great Pyramid obeys so little the development laws of some modern *savants*, that it is the *highest* of all known architectural works up to this day. Man has often tried to erect higher buildings, as in the Cathedral of Cologne, which was to have had spires 6120 inches high; but, either from foundations failing, or some other difficulty, such ambitious structures have invariably stopped short in their progress, leaving the Great Pyramid's height,[1] still, even in these latter days, solitary and supreme over all the architectural works of mankind.

At one time it used to be a fashion, from Pliny amongst pagan Romans, down to very modern society,—to admit the Great Pyramid's superior size, wonder at it somewhat therefore,—and then abuse the king under whom it was supposed to

[1] The following are the heights of some of the principal buildings of the world, chiefly extracted from Fergusson's Architecture :—

	British Inches.
Great Pyramid, originally, . . .	5819 ± 16
Strasburg Cathedral, ...	5616
St. Stephen's, Vienna, . . .	5292
St. Peter's, Rome, ...	5184
Amiens Cathedral, ...	5088
Salisbury, ...	4848
Friburg, ...	4620
St. Paul's, London, ...	4332

have been erected, as the most oppressive of tyrants; and so profligate in expending the national wealth, as to have brought ruin on his country and well-deserved execrations on his own name.

A new system, however, was illustrated last winter,—when an untravelled medical gentleman, at a meeting of the Royal Society of Edinburgh, scoffed very bitterly at both the architecture, and the cost, of the Great Pyramid. Its architecture was so poor, he said, as not to require an architect, only a mason; and any mason in Scotland could build as good a pyramid as the Great Pyramid. While as to its expense, he had been assured by an architectural friend, that it would not cost more than a million of pounds sterling—a sum which had been exceeded two or three times, both by the Taje Mahal and other great Mohammedan tombs in India. Exceeded too many times, as all the world knows, by our Crimean War, whose good results have already vanished into thin air; eaten up in a few months by the mere expense of transports for the Abyssinian expedition—and smoked away by our nation, in the form of tobacco, nearly three times every year.

Whether the building of the Great Pyramid is rightly to be inveighed against for the large, or ridiculed for the small, sum it cost,—we do not pretend to say.[1] 'Our own minute examination of the building in 1865 A.D., remarkably impressed us with the fact, that an undeviating and most considerate *economy*, both of the best workmanship and the finer kind of material, had ruled throughout the building. But that a building, which could not be otherwise than expensive, if it required, for noble, intellectual, and permanent purposes, to represent, amongst many other things, that particular size and mass which we now find it is so eminently doing.

[1] 'Aristotle (*Polit.* vii. 11) considers them (the Pyramids of 'Egypt) merely the result of great labour, displaying the power of 'kings, and the misery inflicted on the people; which Pliny has 're-echoed by calling them an idle and silly display of royal wealth 'and vanity (xxxvi. 12). Later writers have repeated this, without 'even knowing the object they were built for, and it would be un-'just to suppose them merely monumental.'—Sir Gardner Wilkinson, at p. 207 of Rawlinson's *Herodotus*, vol. ii.

INTELLECTUALITY OF THE GREAT PYRAMID—*continued.*

CHAPTER 25.

ITS POSITION.

Latitude.

IN furtherance of its general system of symbolization,[1] the Great Pyramid requires to stand in, or very near to, the latitude of 30 degrees; so as to have the pole of the sky at a certain height above the northern horizon; and also to be on that dividing parallel of the northern hemisphere of the earth, which has an equal amount of the terrestrial semi-surface to the north and south of it.

By careful measures, with a powerful angular instrument, in 1865 A.D., the Great Pyramid's base centre was found to be, not exactly in, but only 1′ 12″ south of, the parallel of 30°; while there

[1] *Life and Work at Great Pyramid*, by C. Piazzi Smyth, vol. iii. pp. 40-52 and 198.

were constructive indications, that the site had been carried as far northward, or towards the parallel of 30° even, as the shape of the hill on which the Great Pyramid is built, would possibly allow; and that building is the closest to the position of all the Pyramids.

Orientation.

Again, the azimuthal arrangement of the Great Pyramid on its parallel of latitude, whatever that may be, or in other words the orientation, *i.e.*, astronomical trending, of the four sides of its base, whether to the cardinal points or otherwise, —is a feature of which a vast deal may be made by an intellectual designer for scientific purposes.

Any one can find out the north, south, east, and west points of the horizon within a few degrees; but astronomy, and sometimes of a high practical order, is necessary to ascertain them to minutes and seconds. Every modern astronomical observatory requires to be built pretty closely in these cardinal directions; and the instruments inside, make up in their still exacter positions, for any residual imperfection of the masonry.

In Tycho Brahe's time (1577 A.D.), when astronomy was far less perfect than now, and rather more trust was placed on the truth of the walls; those parts of his celebrated observatory of Uraniburg, were thought sufficiently close to the truth, when the errors did not exceed 18′. But the sides of the base of the Great Pyramid, built 4000 years previously, are now found, by comparing their 'socket' lines with the Pole-star, duly reduced to the pole by astronomical methods, to be only 4′ 35″ in error.[1]

All working astronomers, and all other men too, who know the difficulty of making near approaches to accuracy in astronomical observations, and especially in large practical works founded thereon,—were delighted with this proof of *more* than respectable performance in the primeval building. For it was a proof showing, as they have expressed the idea,—that astronomy as well as geometry had presided over the construction of the Great Pyramid.

But Sir James Y. Simpson, Bart. and M.D., assuming a position far above all astronomers,

[1] *Life and Work at Great Pyramid*, by C. Piazzi Smyth, vol. iii. p. 109.

recently inveighed, before the Royal Society of Edinburgh, with a peculiar vigour of diction, against the whole practical astronomy of the ancient Great Pyramid builders. Such a proceeding, by such a man, and before such an audience, is important testimony for the guiding of the public, whether to good or to evil; and was thus reported by the *Scotsman* newspaper next morning:—

'He,' Sir James Y. Simpson, 'did not believe one word 'about the Pyramids having been built on astronomical principles. He had no faith whatever in what was called the 'orientation of the Pyramids; they might as well tell him 'that the Register Office and the College were arranged on a 'similar principle.'

And then, according to the *Edinburgh Evening Courant's* report, concluded with—

'It was a spurious archæology and astronomy that fed on 'such fancies and romances.'

As I have nothing more to add to, and nothing to alter in, the hard facts of astronomical observations at the Pyramid, which are printed at length, and with the chief steps of calculation, in *Life and Work*, vol. ii.,—and speak for themselves to every astronomer,—I turn to the Register Office and the College of Edinburgh, quoted by the self-constituted judge, as being just as truly oriented astronomically as the Great Pyramid.

To any one who looks over a map of Edinburgh, it will be seen immediately, that both these buildings (which are but modern structures) are erected conformably to the streets which pass their fronts; and have therefore the same orientation, whatever that may be. The Great Pyramid, on the contrary, being built out in the Libyan desert, and as the first of all buildings in that neighbourhood,—had no help from previously-existing streets to decide its orientation; and yet it attained to $4' 35'' \pm 1'$ perhaps, of astronomical truth. How close then do Sir J. Y. Simpson's model buildings in Edinburgh come to the same astronomical meridian position?

An ordinary Edinburgh map, indicates that approach to be somewhere about 16 degrees: which implies an error, nearly two hundred times larger than what was committed at the Great Pyramid in its primeval day.

Well did Hugh Miller show the importance of any one intellectual arrangement being *several times*, though only a *dozen* times, better than another,—after this manner.

When the proportion of brain to spinal cord in the early geological days of the earth, amounted in any individual animal only to 2·1, hardly anything but fishes existed. When it amounted to 2·5, there were reptiles. When 3·1, there were birds. When 4·1, there were mammals. But when the proportion rose to 23·1, then, as he eloquently expressed it, 'reasoning and calculating man had appeared on the scene.'

Surely then, if we find an astronomical result at the Great Pyramid, not twelve times only, but actually two hundred times better, than at two rivally quoted and professed modern architectural examples,—it is not over much to claim, that there must have been *some* mental intention, and *some* exertion of intellect, at the former well-placed building.

Classical Illustration.

But we may also test the Great Pyramid, under our present head, against those intellectual knights of the ancient world, the Greeks, in their best days.

In the goodly volume on 'Astronomy' published by the Clarendon Press at Oxford, in 1867,

—we read, at p. 463, 'Pytheas of Marseilles, 330 'B.C., first noticed that the so-called Pole-star was 'not situated exactly at the pole.' And as we find by calculation, that the nearest easily visible star to the pole, or in fact the then polar star, must have been some 6° from the pole,—why, the immediate predecessors of Pytheas, or the Greeks at the very height of their civilization,—must have been labouring under an uncertainty of ± 6° at least, in finding the cardinal points by the Pole-star.

Yet the builders of the Great Pyramid, though living 1800 years earlier, had placed their enormous practical structure, within a seventieth part of the error and uncertainty attending the long subsequent science of the wisest Athenian sages; and the building stands in its place to this day, testifying contemporaneously to that fact with far more certainty, than we know any other ancient historical facts whatever.

INTELLECTUALITY OF THE GREAT PYRAMID—*continued.*

CHAPTER 26.

ITS WEIGHT.

So nearly solid a body as the Great Pyramid, must evidently be very weighty ; and it is also evident, that the designer might, within certain limits, make the building of whatever weight he pleased, without too altering any of its other arrangements as to shape, size, or position,—and this merely by constructing the bulk of it, of a heavier, or lighter stone.

Now at the Great Pyramid the prevailing stone of construction is remarkably light ; far lighter, bulk for bulk, than granite ; considerably lighter even than what goes by the name of 'common stone' in most of our tables of specific gravity. Was there anything, then, intellectually gained, by using so anomalously light a stone ? Yes, we may now answer, after having made the calculations. For the employment of any other known stone, or the building being of any other shape or size than it is, or with much more hollow space inside,—would have prevented the Great Pyramid's weight from forming, the very round and even decimal fraction which it appears to do, of the weight of the whole earth!

This fraction is the 1-1000,000,000,000,000th, part; but whose denominator may be more neatly expressed as, $10^{5 \times 3}$; numbers which remind both of the shape of the Pyramid, and the dimensions which are necessary to express solidity, a chief element of weight. Three sorts of material also, of different specific gravities, have to be taken into account, in acccordance with the varying[1] amounts to which they occur in the Great Pyramid. The calculation can hardly be any-

---

		Specific gravity in terms of earth's specific gravity=1.
[1] Granite, an inconsiderable amount, . .	=	0·479
Nummulite limestone, nearly $\frac{14}{15}$ths of the whole,*	=	0·412
Mokattam stone for casing and linings, $\frac{1}{15}$th of whole. See vol. iii. p. 177 of *Life and Work*,	=	0·367
Common stone of the usual engineers' tables,	=	0·442
Basalt of Egypt, ...	=	0·500

* Determined by myself on a fragment of 11·5 lbs. and broken out of the heart of a much larger mass.—December 1867. C. P. S.

thing but an approximate one; and the accordance found, or within $\frac{1}{200}$th of the whole, was both more than had been expected, and more than some of the data,—as the real specific gravity of a great part of the internal material,—can be certainly known to be. And yet, on communicating the results up to this point, to Mr. W. Petrie in London,—he found, that by merely computing the bulk of the earth (with specific gravity assumed as 5·7 that of water) with more attention to small corrections, the accordance was improved many fold.

Referring the reader, however, to Appendix 4, where the numerical particulars are given just as they have thus far turned out, and for whatever they are worth,—we advance to one more general characteristic of the entire Pyramid.

INTELLECTUALITY OF THE GREAT PYRAMID—*continued.*

CHAPTER 27.

ITS HEAT.

STANDING as the Great Pyramid practically does, on the latitude line which equally divides the surface of the northern hemisphere of the world, as a sphere, between the pole and equator,—there is at once a close approach of its climate to being the mean-climate of all the earth's surface, mile for mile; or at least of every individual particle of its area, duly summed up, in every navigable sea and habitable land.

The latter restriction, by throwing out of consideration the small circle of unvisitable circumpolar frozen region,—would have the effect of slightly raising the mean temperature otherwise attributable to the world; and which temperature might otherwise be rather below what is experienced for the air at the Great Pyramid; and much below what is experienced in that part of the parallel of 30° of latitude, where the line passes through the valley of the Nile. This large amount of difference for the Egyptian valley, is nearly, if not quite, removed at the Great Pyramid, by its elevation on a hill top, 2580 British inches above the sea-level. Hence, therefore, it comes just within the range of residual errors of modern science,—whether the mean annual temperature of the air at the Great Pyramid, does, or does not, coincide with the same important temperature element for the entire surface of the earth; and also, whether that heat quantity does not amount to a peculiar division of temperature especially appropriate for the Great Pyramid; viz., a temperature of one-fifth; or one-fifth the space between the boiling and freezing points of water.

The actual temperature observed by myself inside the Great Pyramid, is nearly 6° Fahrenheit too high for the above conclusion; but under circumstances of anomalous temperature disturbance, which might unduly raise a thermometer to that amount; while the deductions from a meteorological journal kept for four months outside the Pyramid,[1] seem to make the air climate come within only 1° Fahrenheit of the expected theoretical temperature-point.

[1] *Life and Work at Great Pyramid*, by C. Piazzi Smyth, vol. ii. pp. 205-270; and vol. iii. pp. 178-202.

INTELLECTUALITY OF THE GREAT PYRAMID—*continued.*

CHAPTER 28.

UNITS AND STANDARDS OF G. PYRAMID MEASURE.

Linear.

HAVING thus briefly touched on each of the five general qualities of the whole Great Pyramid building, viz., *shape, size, position, weight,* and *heat*;—in every one of which traces, and even *consilient* traces, of intention and purpose, highly intellectual have been found,—we turn now to some of that structure's minor arrangements; and shall continue to pursue therein, the same unexceptionably contemporary method of testing the parts of the ancient building concerned, by the most rigorous of recent measures yet procured,[1]

[1] We trust that our readers will always understand, that the best measures procured as yet, are by no means the best procurable in the present state of modern science; if only Government, or the people, would be at the *expense* of applying the more accurate

and weighing them against the results of modern science applied to the earth and the sky.

The absolute size of any π-proportioned Pyramid, having once been stated with reference to its vertical axis,—the sizes of any other parts of the structure, are of course mere mathematical necessary consequences of the two data already laid down; and we need not expect to find by their means, any more intentional coincidences with nature. But, if we measure those residual Pyramid lines in special *units* or *standards*, something new or worthy of notice may come to light; due care being always taken, that some good reason in nature, or expression characteristic of pyramid numbers, shall be represented by the absolute size of such units or standards.

Formation of Units and Standards.

Instructed then now, by what many and various predecessors have gradually ascertained,—we may put the case shortly thus,—

The axis of rotation of the earth, is, for many

methods which are well known to many experts; and thereby improving our knowledge of some of the facts of the Great Pyramid, to a thousandth of their present range of uncertainty.

physical and metaphysical reasons, incomparably the best terrestrial reference that can be employed for linear measure.[1] Let that axis therefore be

[1] In my two former works, *Our Inheritance in the Great Pyramid*, and *Life and Work* thereat,—I have alluded abundantly to what the late John Taylor, Sir John Herschel, and others subsequently to them, have written on this point. But, as a sentence on p. 100, in the introduction to Callet's Logarithms (Paris, 1795), has been since then pointed out, as containing the same idea,—I gladly annex a translation of the chief part of the passage. A striking passage too, to have been published in a country, all whose metrological efforts since then, have been directed towards the propagation of a standard, founded, not as M. Callet would recommend for all nations, on the *axis*, but on part of the *surface*, of the earth. For this, is how that distinguished geometer expresses himself :—

'Men have searched a long time for a universal measure : they 'desired that it should be afforded by Nature, and Nature has offered 'one to which they have not paid any attention. It was, however, 'natural to say—*We wish for a measure which shall serve the whole 'earth ;* well then ! let us take the axis of the earth. It is a 'straight line ; it is the distance from one pole to another ; let us 'compare then, to this distance, all those that we wish to know. 'If men take up this view, the length of the ten-millionth part of 'the earth's axis being called the standard, they will find for the 'mean length of this standard 3 feet 11 inches French, to within '$\frac{1}{100}$th of a line.'

It has also been said, that in a work of *Cassini's*, published in Amsterdam in 1723, there is a recommendation of the polar axis of the earth as a reference for linear measure. I have not yet been able to meet with the book in question ; but a friend who has kindly looked it up at the British Museum assures me, that there is no mention of the *polar axis* for such purpose ; only of a general *diameter*, anywhere between the pole and equator. Indeed, *savants* were in doubt in 1723, whether the earth was flattened or protuberant at the poles. But as Cassini proposed to divide such general semi-diameter of the earth into 10^7 parts, and showed certain trigonometrical advantages to flow therefrom, he was unconsciously testifying to one feature in the principle of the 'sacred cubit,' as the 10^{-7th} part of the earth's polar axis.

subdivided, illustratively of the pyramid numbers, into five hundred millions of equal parts ; and let each of these be considered a *unit* or *inch* of Pyramid linear measure.

And let 5 × 5, or 25 of these units, form the *standard* or *cubit* of the same measure ; having thence the property of being the ten-millionth part of the earth's polar semi-axis. Or, in other words, 10^7 of such standards will measure the shortest distance from the centre, to the surface of the earth, at either pole.

This is in so far an ideal arrangement only ; but let us next try to get a practical knowledge of the length of such units and standards by reference to known earth-measures.

The best, latest, and most comprehensive determinations of the size and shape of the earth,—show, that its polar axis is, in length, somewhere between 500,482,296, and 500,522,904 British inches ;[1] and we cannot in the present state of science be certain of coming much nearer. If

[1] See Captain A. R. Clarke, R.E. ; in the Ordnance Survey volume, entitled *Comparisons of the Standards of Length, of England, France, Belgium, India, and Australia.*

See also *Life and Work at the Great Pyramid*, by C. Piazzi Smyth, vol. ii. p. 450.

therefore, having reference to this residual error, we subdivide the mean of the two quantities given by five hundred millions,—we obtain, for the length of one of the Pyramid units, in terms of British inches, 1·00101, with an uncertainty of ± ·00004.

While the Pyramid standard or cubit, being 25 of these units, must necessarily measure in British inches, 25·025, with an uncertainty of ±·001.

These results, in themselves, we presume, will be objected to by no one; though some person may very properly inquire, what material connexion has either the theoretical unit or theoretical standard with the actual Great Pyramid?

To this, we answer, as a beginning,—that the said standard goes into the length of one side of the Pyramid's base (supposed to be 9142 British inches[1]), 365·3 times. Or, so nearly the number of days and portions of a day contained in a year, that we are inclined to suspect that the exact and true quantity may have been intended. We are prevented, indeed, from pushing the examination closer at this point just now, by the large amount of uncertainty hanging over even the best of the

[1] See Chapter 24, p. 205.

base-side measures yet made at the Great Pyramid, by all modern nations.[1] But we may remark, in connexion with another feature of the building, that there are *four* similar sides to the base; which sides, expressed in terms of the Pyramid cubit or standard (if perfect for one of them), will cause the whole four to represent numerically, the number of years in which the fraction of a day at the end of a year, is practically restored to a

[1] The length of base-side, it will be remembered, is still uncertain, according to my estimate, by the amount of ± 25 inches. These limits quite allow of Mr. Petrie's conclusion from all the measures, of 9165 British inches, being very probably the actual fact of the building. If that be so,—the Pyramid cubit of 25·025 British inches, goes 366·24 times into the base-side; and all that we are reasoning on touching *solar* days, applies really to the earth's *sidereal* days, or times of its revolution round its axis, as measured by the recurrence of a distant fixed star to the meridian of any place. This quantity would be, under many points of view, the higher scientific conception to have expressed in the building; but the sober doubt immediately arises,—'is it so expressed?'

The answer cannot be rendered with practical certainty, until some Government, or nation, has ordered the base of the Great Pyramid to be measured with the same care (and for that end at something like the same *expense*) as the usual base-lines employed in most of our large trigonometrical surveys. The operation, when performed (say to the accuracy of ± 1-10th of an inch, or less, on a run of 9000 inches), *might* of course show, that the Great Pyramid base-side was neither 365·24, nor 366·24 times, 25·025 British inches,—but some awkward, unscientific, and non-symbolizing number; and such a proving of non-intellectuality against the building's original designer, would delight some persons extremely. Would that it might happily prevail with those who are in power in high places, to have the work performed forthwith, and by three independent sets of observers, instruments, and processes!

whole number: a knowledge of which fraction and which cycle, is a necessary beginning and foundation for all the chronological reckonings of mankind. And this very admirable result appears only, when the base-side is measured by a standard, whose length is founded on an even Pyramid portion of that axis of the earth, whose existence is a function, and necessary accompaniment, of the diurnal rotation itself.

Such coincidence working into coincidence, and with the effect of developing yet further those earth and solar relations already commenced in other portions of the building,—cannot be merely accidental.

The Pyramid cubit, which brings out so many of these things when applied to the Pyramid,—is no doubt a very scientific standard in itself; and far above the human science of that day, or the following 3800 years of the world's history, to have obtained from nature; and is not known to have been in use among any Gentile nations. But then Sir Isaac Newton has shown long since, that a length very much like that of the Pyramid cubit,—certainly on the mean of all his data, duly weighted, much closer to it than the amount of his probable error,—was the sacred cubit of the Israelites; *i.e.*, a cubit known to, and appreciated by that peculiar people, as sacred to them, before they entered Egypt under Jacob, and after they left it under Moses.[1] A cubit too, supposed to have been given to them by Divine inspiration; used by them for none but sacred purposes; and distinguishable only, but abundantly and easily, from the profane cubits of Egyptians, Babylonians, and all other Pagan nations, by its length, without necessary reference to anything else.

My own researches also, on direct Biblical notices of the 'Ark of the Covenant,'—the most solemn portion of the contents of the Tabernacle, —tend strongly to confirm Sir Isaac Newton's views;[2] and perhaps even to settle, far within the present residual imperfections of our knowledge of many portions of the data,—that the scientific cubit of the Great Pyramid, and the sacred cubit of the Hebrews, defined to them by inspiration, —were identical as measures of length.

[1] See Sir Isaac Newton's 'Dissertation on Cubits,' in Professor Greaves' Works, collected by Dr. Birch.

See also *Life and Work at the Great Pyramid*, vol. ii. pp. 341, 366; and the Errata thereto, printed at the end of this book.

[2] *Life and Work at the Great Pyramid*, vol. ii. p. 465.

Counter Opinions.

Amongst those who dislike the primeval testimony of the Great Pyramid—such a result as that just mentioned, was not likely to pass unquestioned, unattacked, and even, as the discussion has recently been carried on, unbattered against. Nor is there any objection to be taken against gentlemen satisfying their individual and critical judgments in that manner, if they will,—so long only as they use fair arguments,—honest methods, and tests allowed by all scientific men in every other branch of science.

But some of the arguments recently employed before the Royal Society of Edinburgh are doubtful indeed;[1] *ex gr.*—

1*st.* The insinuation that it was very '*strange,*' —that Professor Smyth, though in Egypt for the purpose of measuring the Pyramid, never measured from socket to socket.

Yet the reasons are printed and published plainly enough in *Life and Work;* and show that Sir J. Y. Simpson might as well taunt Consul Cameron and the Abyssinian captives with not freeing

[1] See Appendix 2.

themselves, in spite of opposition and difficulties, which other persons have thought removable only by sending a large army, at an expense of several millions of money, to their assistance ; as inquire, with well-feigned astonishment, before a public meeting, why the Astronomer-Royal of Scotland, in his mere private capacity, did not perform a task, whose proper and accurate execution is impossible for any but the wealthiest of men, or most powerful of nations ; and has never yet been performed by any country or people. The said Astronomer had not a farthing of public, and very little private, money when in Egypt ;—and, although anxious to do whatever he individually could, towards measuring parts and features of the Pyramid already laid bare for measurement, he had no more notion of undertaking extensive works which are the duty of Government, than the wealthy Baronet is believed to have, of employing his private means for paying off the national debt.

2*d.* 'Professor Smyth lays down,'—so says the same writer, 'as the measurement of one side of 'the base, the figures 9142 inches (which no one 'ever found it to be), because that figure will be 'the result of multiplying the number of the 'days of the year by what he calls the pyramidal 'cubit, made up of twenty-five mythical pyra-'midal inches.'

That the *bad* reason given above, was not that which influenced Professor Smyth in considering the most probable length of a base-side of the Great Pyramid to be, 9142 British inches,—must be apparent enough to any one who will read *Life and Work at the Great Pyramid*, vol. iii. pp. 123 to 143; for there, are *all* the measures the author could collect of *all* persons who had measured between the only fiducial points, viz., the corner sockets. All such are given in full, simply and independently at first; and then combined together according to a stated principle, for arriving at the most probable mean; and what other plan could have been taken in the cause of honesty and uprightness?

The critic did indeed so far explain in his discourse, and did it with some emphasis, as though denouncing a moral delinquency of the gravest order,—that the figure 9142 which Professor Smyth had laid down as the base-side length, was one '*which no one* (observer?) *ever found it to be.*'

Precisely so, because a mean of several varying observations, seldom falls on any one in particular; and in this case three of the observers had recorded results all above 9160, while a fourth one was nearer to 9110. Wherefore, when the simple mean of all their observations is taken, how can it possibly fall anywhere else, than in that unfortunate gap which divides the results of one of these observers so widely from those of the other three?

Perhaps indeed that fourth observer's result was rated at too high a worth;—and ought to have received only a very small proportional weight to any of the others; but even if it had been actually rejected, and if the three results above 9160 had remained the only ones to be used,—the *mean* amount of *them*, might have also fallen not on any one in particular, but between two;—and therefore the moral accusation could have been still brought up, against any unfortunate computer,—if there be any truth whatever in its principle—*i.e.*, that the final quantity he had arrived at, 'was a figure which no observer *had ever found it to be.*'

In such case, however, it is not Professor Smyth

only that is being attacked,—it is the whole principles of arithmetic, and all those who make use of them, high and low.

3*d*. A more serious part still, was performed by the eminent Baronet, when he stated, *inter alia*,

'that if Sir Isaac Newton's standard of the "sacred cubit" 'be true, and if the polar axis of the earth be, as fondly and 'theoretically imagined by Professor Smyth, 500,000,000 'pyramidal inches, then the brim of the lecturer's hat being 'exactly one-half of Sir Isaac Newton's "sacred cubit," that 'brim was in measurement one 20,000,000th of the earth's 'polar axis.'

The said hat, and a scale to go through the form of measuring it with, had been laid on the titled lecturer's desk by his assistant, before the discourse began;—and when the right time came, the pretended measurement was performed on the hat, in place of the sacred cubit of Moses as determined by Sir Isaac Newton; and performed with so much unction of manner and look, as to be received with cheers by the large and learned audience (with a few exceptions), in the hall of the Royal Society of Edinburgh.

Ridicule has always been held to be a powerful weapon, but exceedingly dangerous; especially when employed against sacred things. Because, although these may stand their ground unmoved in themselves,—the souls of many good and worthy, but not very strong-minded brethren, amongst the spectators, may be led away. They *were* led away on this occasion,—and Sir J. Y. Simpson's position is not an enviable one, if he drew them from the truth.

To take the matter, however, only in its less serious light,—after higher minds, whether in the primal ages of the world, or more recently, have found out what the length of the 10^{-7th} part of the earth's polar semi-axis is,—there remains no difficulty in *any one* taking up some dirty or ignoble portion of matter, and showing that it is approximately the half, third, fourth, or any other simple fraction of the same 10^{-7th} part. But does such a mere jester, prove that he is entitled thereby, to scorn in learned halls, the labours of those higher minds which had preceded him, and treat them with studious insult?

A nearly parallel case, in much of its principles, with the Pyramid or 'sacred cubit,' is offered more within our immediate historical records, by the origination of the French standard of length, *the metre.* In so far as that is, or was intended to be, an earth-commensurable measure (viz. the ten-millionth of the length of a quadrant of the meridian, extending from the North Pole to the Equator), it should be equally amenable to being laughed at, insulted, degraded, and abolished,—because some one should chance to show that a hat, or a shoe, or any other absurd portion of garment or vesture,—should be nearly an even fraction of it, and therefore some mentionable fraction of the earth's meridional quadrant.

Yet if a gentleman who has never performed any notable problem in exact researches, should appear in the Imperial Academy of Sciences in Paris,—where for eighty years the greatest mathematicians and geodesists of France have spent their best labours in perfecting the methods, both theoretical and practical, by which they have arrived at last at a close approximation to what the length of their metre (or the ten-millionth of the earth's quadrant) ought to be;—and if he should there gratuitously insult all the members, on an important public occasion, with his hat-trick, applied to the *metre*,—will they thereby be put out of conceit with their national standard which they now prize so much, and hope to see adopted by every nation of the earth;—or will their quick susceptibilities be aroused? Sir J. Y. Simpson has been very bold against what many men believe, and there are some proofs to show, must have been a divinely originated measure; but will he exhibit equal bravery against the French nation, and their similarly earth-commensurable standard? We do not advise him to try; for when the national pride of our Gallic neighbours is aroused, by a heartless insult in an important matter of their national science and glory,—they are neither long-suffering nor merciful.

Still another reason was brought forward by that same gentleman at the Royal Society, Edinburgh, against the Great Pyramid inch and 'sacred cubit.'

'It was too late now,' he implied, 'to be 'talking about a new standard measure for this 'country, because Mr. Whitworth of Manchester 'had adopted another size of inch, and had been 'making for several years past extraordinarily 'accurate copies thereof as standard measures.' Two of these, a cylinder, and a hole in a block, each said to be true to 1-10,000th of an inch, were exhibited, and seemed to be considered final.

Yet Mr. Whitworth's gauges of one or many inches are not adopted for the standard measure of the country by Government, who have their own linear standards, in a perfectly different shape;—nor used anywhere, except for certain practical purposes in engineering workshops;—neither do they lead, on Mr. Whitworth's plans, to any convenient arrangement or good system of greater length-measures than inches: while they are totally unconnected with all other classes or kinds of measure, as weight, capacity, heat, and time;—and are, in themselves, without that inestimable merit, which the sacred measure of old and the scientific societies of modern time, alike appreciate and appropriate for its thought-elevating influences to man; viz., even, and direct,—earth-commensurability.

Had any one said that it was too late to talk of introducing the Pyramid measures, because the French *metrical* system has already established itself amongst us,—there would have been statistical facts on his side. For, poor Mr. Whitworth's inches, as well as our own national hereditary measures, are all going down before the French metrical system. Already there is a permissive bill for the latter's legal adoption in certain cases, passed by Parliament; every meeting of the British Association for several years past, has re-echoed with the demands of several sections to have the metre taught in all schools and employed in all Government offices; and not a few of the authors of papers in the Philosophical Transactions of London, employ French weights and measures; even when dealing with purely British investigations for British purposes.

To assure the Royal Society of Edinburgh, therefore, that they had nothing to do, in questions of metrological arrangement, but trust implicitly to Mr. Whitworth's iron gauges,—was very much like crying 'peace, peace, when there is no peace,' and when the enemy is actually in possession of the gates. There, indeed, he is now, and from that position, only a most vigorous resuscitation of the true national feeling, will drive him; for it has long been a settled purpose in France, to leave no stone unturned, in order to prevail on other countries,—but chiefest of them all Great Britain,—to adopt her metrical system;—and there is a very numerous pro-French party, ever active and eagerly pushing, amongst ourselves.

Their great battle-cry too, is, the scientific completeness of the French metrical system, through every range of subject requiring mensuration. They will certainly therefore not be content with the simple, solitary, barren, and unscientific mere thickness-gauges of Mr. Whitworth. The metrological requirements of the world, in the present advanced day, are for more science, system, and order; and the world finds more of all these in the French metrical system, than in any other system, or no-system, of weights and measures, that has ever been known to exist amongst men in all historical time. Yet what, if there should now come forth from the Great Pyramid (which has already given us, in the ten-millionth of the polar axis, a better earth-reference than that of the French metre), yea, what if there should come forth from thence also, a *better* arranged whole system, broader even and more scientific, for all weights and measures,—as well as bearing on its front the dignity of primal time, the impress of Divine inspiration, and some strange family traits of, and general resemblance to, our own national and hereditary system; which our nation had acquired somehow, in a manner which no one has yet explained, long before we emerged into the light of history.

A higher Earth-reference.

Already the Great Pyramid linear system, while still preserving its cubit reference by 10^7 to the earth's polar semi-axis,—has yielded to the further researches of William Petrie,—the knowledge of a far grander reference yet in nature, and of a more comprehensive kind in science;—but expressed in similarly round and even decimal repetitions of Pyramid numbers and of Pyramid units of linear measure.

Of all the higher mechanical properties, and movements of the earth, none is perhaps so mighty in itself, so terrific in its momentum, so all-important to the safety of humankind to have preserved; and so constantly and uniformly kept up,—as the onward movement of the earth in its orbit round the sun. Making its headway at the rate of more than 65,530 miles an hour, what an instrument does it become for measuring the scale of creation! By dint of great labour and expense, astronomers may travel to nearly opposite sides of the globe,—as to the respective Observatories of Greenwich and the Cape of Good Hope; —and by, in that manner, putting five or six thousand miles between them, they may determine the parallax of a heavenly body, if not very far removed. But if we choose rather to trust to the earth's orbital motion, we shall have gained in five minutes the same amount of space, or two-thirds of the earth's breadth, on our former position. And if we merely delay repeating our observations for a fortnight, no less than twenty millions of miles will have been traversed, in easy company too, with all our friends, and their houses, and thoughts, and fancies.

When the old Roman poet wished to give an idea of the most rapid attempt to escape from immaterial pursuers,—he placed his hero on a horse, a round-sided fat Roman horse, and set him galloping along the Appian Way at the rate of eight or nine miles an hour; and then after such a magnificent course as that, the belted knight turns him about, and lo—black Care is still seated behind his back.

But that author was a Pagan poet, and his writings were flights of the imagination; here we have only to do with plain facts of measurement and sober realities of creation; or with the world and all that is therein and thereon flying forward together at a rate of 10,000 times the speed of cavalry. And therewith comes the practical question, what given length of such path is swept over by the earth, in that special period or standard for the measurement of time (so admirably uniform in itself, and so intensely important in ruling the affairs of men),—as the interval represented by the whole earth turning upon its polar axis; and bringing a day of toil and a night of rest to all the wearied nations of mankind.

The answer is (as given by Mr. Petrie), if you employ Pyramid *units* or inches, of linear measure, you may tell off that mighty standard of space and force in a round decimal number, or by 10^{7+4}, *i.e.*, by 100,000,000,000 Pyramid inches.

The precise proof of this formula, must inevitably wait the year 1882, for both the more correct determination of the sun's parallax, and also, a proper and accurate measure of the Great Pyramid's real size. But meanwhile, taking it in this mere first approximate way, 5819 British inches for the Pyramid's height, reduced to Pyramid inches, becomes 5813·2; which quantity $\times 10^9$, gives our present best determination of the earth's radius rector.

Then the circumference of the earth's mean orbit $= 5813{\cdot}2 \times 10^9 \times 2\pi = 36{,}525{,}430{,}000{,}000$;[1] and this quantity, divided by the number of solar days in a sidereal revolution of the earth, or 365·25636, yields a little under the 10^{7+4}, or 99,999,400,000 nearly.

Pyramid inches therefore, do indeed promise well to measure off this wondrous and exalted standard in nature, in even and round decimal numbers, with a very small proportional error only, if not perfect exactness. While either British yards, or French metres, would be inconceivably rough and untoward.

[1] The seven last places are entered as 0s, not because they are that, even; but because two previous places of numbers being already erroneous, it does not signify at present what these last are; though we must have something to show the number of places of figures in the whole expression, and 0s are the most convenient for this purpose. As, for example, in the table of tons weights of the planets, p. 39 of G. F. Chambers's *Descriptive Astronomy*, in the Clarendon Press series; Oxford, 1867.

INTELLECTUALITY OF GREAT PYRAMID—*continued.*

CHAPTER 29.

UNITS AND STANDARDS OF GREAT PYRAMID MEASURE—*continued.*

Weight and Capacity.

IN the interior of the Great Pyramid, nearly in the centre of its substance and weight, is a certain chamber, known vulgarly as the King's Chamber; and the only piece of furniture therein, is a hollow, empty box, or open trough of hard stone.

Polished red granite in large slabs the floor, the walls, and the ceiling of the room; polished red granite also, of a rather darker kind, the said uninscribed, unadorned, and lidless box.[1] By some called sarcophagus, the said box has been thought by that school, to have entombed the royal founder of the Great Pyramid, and to be merely a coffin; by

[1] The material has been by several writers termed *porphyry*, myself included, *before* going to Egypt. But from the moment of seeing it there, I was convinced of the error.

others named the coffer, it is thought to have been a large standard measure of capacity and weight.

John Taylor was of the latter idea; and some other persons, after having had long practical acquaintance with the thing itself, as well as with ordinary Egyptian sarcophagi, came to be *generally* of his opinion.

That is, they allowed that there was much in his view; but that that view was not *therefore* altogether exclusive of the other. Some laying up of mortal remains in honour, may therefore, according to such opinions, once have taken place, in the long rectangular, granite box. But, if ever such remains were so deposited, not only has every trace of them vanished long since, and there is no inscription of any kind to testify for them,—but, in their place, distinct footsteps of geometry and physical science have been discovered.

Now, such scientific features would evidently not have been necessary, had that vessel been intended *only* for a sarcophagus; but were eminently adapted for giving it a learned character, perhaps symbolic of, if not actually representing, a standard measure of weight and capacity; a standard, too, as careful observations and mensurations have since indicated, of a most high and noble earth-reference kind.

By a subsequent series of investigations, moreover, the cubical contents of the said vessel have been found to be a close representation of those of the sacred Ark of the Covenant, built by Moses under direct inspiration commands of number and measure, recorded in the Scriptures, for the Tabernacle in the Wilderness; or, taking the two box structures according to their respective dates, the latter, *i.e.* the Ark, was in cubic internal capacity, a very reproduction of the coffer in the King's Chamber of the Great Pyramid.[1]

The Attack.

If ever the weak points of a theory were likely to be discovered—and perhaps it may be salutary that every theory shall be tried by the rudest shaking—it was when Sir J. Y. Simpson attacked the metrical theory of the coffer in the King's Chamber of the Great Pyramid of Jeezeh, before the Royal Society of Edinburgh, on the 20th of January 1868. Certainly he spared neither pains nor labour, neither ridicule nor invective, nor

[1] See *Life and Work at the Great Pyramid*, vol. ii. p. 465.

considerable mis-statement of facts, nor quotations from unauthoritative authors, to overwhelm both the coffer and Professor Smyth, one of its advocates for honour, under a storm of utter obloquy.

So that then, when after concluding an oration, of whose length and skill, design and accompaniments, the report in our Appendix 2 gives but a poor idea; and during which the speaker kept before the eyes of his audience a neat and shapely diagram of the coffer, stated in writing to be from Professor Smyth and extracted out of his book, *Our Inheritance* (written *before* he went to Egypt),—when just at that moment of cumulative interest, the lecturer with a practised hand dashed up an intervening screen, and showed his audience a great perspective picture of the coffer, rude, dingy, broken, and apparently tainted with corruption,—as being the real thing itself,—many of the listeners shouted with satisfaction; and Professor Smyth found himself suddenly knocked down violently, by information chiefly taken out of, and much perverted from, his own published book, *Life and Work at the Great Pyramid,* written *after* he had visited that most ancient and remarkable structure.

Following up this successful outburst, there was little thereafter left for the titled rhetorician, but undisturbedly to ride roughshod over the religious beliefs of some other men,—and administer the *coup de grace* in the following words :—

'In short, the marvellous metrological coffer, mathematically 'formed under alleged Divine inspiration as a measure of 'capacity (and, according to M. Jomard, of length also), for 'all men and all nations,—for all time, and particularly for 'these latter profane times,—is in truth, nothing more than 'an old and dilapidated stone coffin.'

The Defence begins.

'Tell us,' said to Professor Smyth, the spokesman of a little knot of members of the Royal Society of Edinburgh, after the public part of this meeting was over, 'tell us,—just this one thing ; 'it is the only one we care to know, and it seems 'to our minds absolutely conclusive. Did the 'Arabian writer, Al Hokm, really mention, as 'stated by the lecturer, that Khaliph Al Mamoon 'found the body of a man deposited, with jewels, 'arms, and golden writing, in the coffer, when 'he broke into the King's Chamber of the Great 'Pyramid ?'

'No doubt he did,' was the answer ; 'but then,

'seeing that the said Ebn Abd Al Hokm (according to the chronological arrangement of the 'Arab authorities by Colonel Howard Vyse and 'Dr. Sprenger, in the former's second volume of '*The Pyramids of Gizeh*), lived six hundred 'years after the death of Al Mamoon,—what sort 'of safe or contemporary authority can Al Hokm 'be, for anything that the Khaliph, when alive, 'did, saw, or got hold of, in a certain dark room 'in the Great Pyramid six centuries previously?'

Al Hokm testified to it, we allow; and to much besides of a necromantic and absolutely impossible nature. But then, who will vouch for Al Hokm? He was a follower of a false prophet, a believer in sorcery, enchantment, and the existence of Jinn; besides being an upholder of a theory as to the Great Pyramid having been built by a certain antediluvian King Saurid, and filled by him chiefly with celestial spheres and figures of the stars; together with the perfumes to be used in their worship.[1] So that Al Hokm is neither a trustworthy writer in general,—nor in this Al Mamoon particular is he a *contemporary*

[1] See in the first volume of Professor Greaves' Works, p. 115, a longer translation from Al Hokm than what appears in Colonel Vyse's book.

authority; and that, by so large a space of time as six hundred years in the darkest of the dark ages; wherefore Ebn Abd Al Hokm can have neither place nor peculiar honour from us in this inquiry.[1]

Why Sir J. Y. Simpson should have picked out so very late an author as Al Hokm, and trusted to merely the last sentence of his long and rambling account of the Great Pyramid implicitly, for what the poor man could be no direct authority for,—it is very difficult to say. Because, if the object was merely to get at the best Arab testimony as to Al Mamoon's proceedings, there are plenty of earlier authorities than Al Hokm; and they have been admirably translated and chronologically set forth in the second volume of Colonel Howard Vyse's *Pyramids of Gizeh.*

Now Khaliph Al Mamoon flourished between 813 and 842 A.D., and may or may not, have made his celebrated entrance into the Great Pyramid

[1] A rumour has now reached us of a manuscript note, just written by a gentleman of the British Museum to say, that Al Hokm was nearly contemporary with Al Mamoon. But while the long-printed authorities of Vyse and Sprenger are not by such a rumour only, instantly proved to be erroneous,—it is believed that this new reference was not in existence at the time of the lecture before the Royal Society above alluded to.

about 830 A.D.; but be that as it may,—the immediately succeeding authors, in Vyse, are silent on the point. Thus neither Abou Masher Jafer Ben Mohammed Balkhi, who wrote in about 890 A.D., nor Ebn Khordadbeh, in 920 A.D., have one word about Al Mamoon, or about any opening of the Pyramid. But when we descend to Masoudi, in 967 A.D., he, after an astonishing amount of romancing on what took place at the building of the Pyramids 300 years before the Flood,—mentions that, not Al Mamoon, but his father, Khaliph Haroun Al Rasheed, attempted to break into the Great Pyramid; and after penetrating 20 cubits, found a vessel containing 1000 coins of the finest gold, each just one ounce in weight, and making up a sum which exactly repaid the cost of his operations; at which, it is added, he greatly wondered.

Down, down in fact along the stream of time must we descend further; and at last, in or near 1170 A.D., or 340 years after Al Mamoon's age, that chief is mentioned by Abou Abd Allah Mohammed Ben Abdurakim Alkaisi; who states that he was *informed* that those who went into the upper parts of the Great Pyramid in the time of Al Mamoon, 'came to a *small passage*, con- 'taining the image of a man in green stone, and 'within that a human body with golden armour,'[1] etc., etc.

Subsequent authors, however, seem to take no notice of this story, though they are fond of attributing to Al Mamoon the finding of the emerald pot full of gold at the end of his quarrying near the entrance-passage; until Al Hokm, in or near 1450 A.D., as already mentioned, gives new circulation to the figure of the man in green stone containing the real man; but now as being found 'in an hollow stone, and that again in a chamber towards the top of the Great Pyramid.'

The Arab authors, then, even of the most favourable kind, are both exceedingly scanty, and a sad long time after date in describing Al Mamoon's finding anyhow or anywhere the body of a man. An earlier author too than Al Hokm, or Shehab Eddin Ahmed Ben Yahya (died 1317 A.D.), though stating that the Great Pyramid was opened by Khaliph Al Mamoon, adds, in opposite

[1] The same notice is given soon after by *Abou Szalt* of Spain, but at the end of some extraordinarily romancing stories of the cabalistical and other wonders inside the Great Pyramid. See pp. 355-357 of vol. ii. of Vyse's *Pyramids of Gizeh*.

testimony to Al Hokm, 'that nothing was dis-'covered as to the motive or time of its con-'struction.' And his own idea seems to be, that each of the Pyramids was dedicated to a star, and that the Sabæans performed pilgrimages to some of them, but especially to the Great one.

Other Arabian authors again, who are also earlier than Al Hokm, and much nearer to the time of Al Mamoon, have described very different findings, but as actually found, by Arabs; thus Abou Hajalah, in 1390 A.D., writes that—

> 'In the days of Ahmed Ben Touloun' (870 A.D.—you may see his mosque in Cairo to this day; and he was, for Egyptian local government, the Mehemet Ali of that early time), 'a party entered the Great Pyramid. They found in one of 'the chambers a goblet of glass of rare colour and texture. 'When they came back they missed one of their party, and 'upon returning to seek him, he came out to them naked, 'and laughing, said, Do not follow or seek for me, and then 'rushed back into the Pyramid. They perceived that he was 'enchanted, and told the circumstance to Ahmed Ben Touloun, 'who forbade that any should enter the Pyramid, and took 'possession of the goblet of glass, which was afterwards 'weighed, and found to be of the same weight empty, and 'when full of water.'

In a similar vein, too, the afore-mentioned Masoudi, in 960 A.D., only 130 years after the times of Khaliph Al Mamoon, but 490 years before Al Hokm,—describing the early discoveries of his countrymen inside the Great Pyramid,—says—'that some of them found in the lowest part of the Pyramid, 'a vase containing a quantity of fluid of an unknown quality. 'The walls of the chamber were composed of small square 'stones of beautiful colours, and a person, having put one of 'those stones in his mouth, was suddenly seized with a pain 'in his ears, which continued until he had replaced it. They 'also discovered, in a large hall, a quantity of golden coins 'put up in columns, every piece of which was of the weight 'of one thousand dinars. They tried to take the money, but 'were not able to move it. In another place they found the 'image of a shiek, made of green stone, sitting upon a sofa, 'and wrapped up in a garment. Before him were statues of 'little boys, whom he was occupied in instructing; they tried 'to take up one of these figures, but they were not able to 'move it. Having proceeded further to a quadrangular space, 'similar to that which they had previously entered, they met 'with the image of a cock, made of precious stones, and 'placed upon a green column. Its eyes enlightened all the 'place; and, upon their arrival it crowed and flapped its 'wings. Continuing their researches, they came to a female 'idol of white stone, and lions of stone on each side, attempting 'to devour her; upon which they took to flight. This occurred in the time of Yerid Ben Abdullah,' supposed, says Colonel Howard Vyse, to have been a king of Egypt.

Dreadfully untrustworthy then as these Mohammedan Arabs show themselves to be, even when they purport to be merely describing what took place near their own times inside the Great Pyramid,—it is no wonder that Professor Greaves, 200 years ago, came to the conclusion—

'Thus far the *Arabians;* which traditions of theirs are 'little better than a romance.'

Or that Colonel Howard Vyse, our next Great Pyramid authority, writes, after a more extensive examination of Arab writers than any one else, either before or since, has made,—

'The only fact which seems to be established by the Eastern 'authors, to whom we have now referred, is the opening of 'the Great Pyramid by Al Mamoon; and even of that, no 'distinct or rational account exists.'[1]

---

[1] In the 52d chapter of Gibbon's *Decline and Fall of the Roman Empire,* Khaliph Al Mamoon receives such abounding praises as the chief mediæval patron of learning,—that we find it exceedingly difficult to correlate them with the above-mentioned Arabian accounts of his own, and his Egyptian subjects', doings at the Great Pyramid. Perhaps he reserved all his learning for the politer society of his chief capital on the banks of the Tigris. Certainly Gibbon writes of him, mostly as the ruler of Bagdad, that he, Al Mamoon (or Almamon), 'completed the designs of his grandfather, 'and invited the Muses from their ancient seats. His ambassadors 'at Constantinople, his agents in Armenia, Syria, and Egypt, col'lected the volumes of Grecian science; at his command they were 'translated by the most skilful interpreters into the Arabic lan'guage; his subjects were exhorted assiduously to peruse these 'instructive writings; and the successor of Mahomet assisted with 'pleasure and modesty at the assemblies and disputations of the 'learned.—They (the Arabs) cultivated with more success the 'sublime science of astronomy, which elevates the mind of man to 'disdain his diminutive planet and momentary existence. The 'costly instruments of observation were supplied by the Caliph

While Sir Gardner Wilkinson adds on the same side, though with exceeding gentleness to those whose posterity he passed many years of his life amongst,—

'The authority of Arab writers is not always to be relied 'on; and it may be doubted whether the body of the King 'was really deposited in the sarcophagus.'[1]

Sir Gardner perhaps also remembering, that Diodorus Siculus, 900 years before Al Mamoon, states that neither of the Kings of both the Great and second Pyramids were buried therein. And Herodotus, 400 years earlier still, said that King Cheops of the Great Pyramid was

'Almamon, and the land of the Chaldæans still afforded the same 'spacious level, the same unclouded horizon. In the plain of Sinaar, 'and a second time in those of Cufa, his mathematicians measured 'a degree of the great circle of the earth, and determined at 24,000 'miles the entire circumference of our globe.'

In a note the eloquent historian adds, 'This degree contains '200,000 royal or Hashemite cubits, which Arabia had derived 'from the sacred and legal practice both of Palestine and Egypt. 'This ancient cubit is repeated 400 times in each basis of the 'Great Pyramid, and seems to indicate the primitive and universal 'measures of the East. See the *Metrologie* of the laborious M. 'Paucton, p. 101-195.'

This very indirect connexion, is the only one which Gibbon gives us, between Al Mamoon and the Great Pyramid. Paucton's error therein has been already exhibited in my *Life and Work*, p. 129, vol. iii.; while Gibbon's mistaken idea of the real length of the *sacred* cubit of the Hebrews, is illustrated at pp. 455, 456, vol. ii. of the same work.

[1] Murray's *Handbook to Egypt* (1867), p. 168.

buried, not in an upper chamber like the King's Chamber now so called, but in a subterranean apartment surrounded by the waters of the Nile.

Even too, had the Arab authors been a more truth-speaking and science-understanding set—the most important fact yet remains; viz., that no matter what the Khaliph Al Mamoon, or any other early Arabian plunderer may have found, or did find, in the King's Chamber,—that would not necessarily show what the original builders had intended it to contain, or did place therein. Because, there is a growing certainty amongst Egyptologists to the effect, that the Arabs had been preceded in their plundering, long ages before Al Mamoon, by the old Egyptians; but by Egyptians of a date far subsequent to the foundation of the Great Pyramid, and having no religious ideas in common therewith. What smaller items of furniture, therefore, the Arabs of 830 A.D. found in the room, if they did find any, depends more on those mediæval Egyptians (probably of 600 B.C.), than on the original founders of the Pyramid, in 2170 B.C.

This matter is unmistakably alluded to by Sir Gardner Wilkinson, in the following words,—

'That both the Pyramids had been opened before the time 'of the Arabs is exceedingly probable, as we find the Egyptians 'themselves had in many instances plundered the tombs of 'Thebes; and the fact of its (the Great Pyramid) having 'been closed again is consistent with experience in other 'places. Belzoni's tomb had been rifled and reclosed, and 'the same is observed in many Theban tombs, when dis'covered by modern excavators.'[1]

Colonel Vyse also writes, in the notes accompanying Mr. Perring's large plates of the Pyramids (p. 2 of Part 2),—

'By whom the Pyramids were first broken into, will pro'bably never be revealed; but, as they were apparently 'entered by the regular passages, their interior construction 'must have been at that time known. It appears, likewise, 'that the Mahometan Caliphs who, according to Arabian 'historians, made the forced passages, must have possessed, 'either by tradition or otherwise, some information respecting 'them.'

And we ourselves too, were unavoidably impressed, on examining the features of the stone at the place where the well-mouth enters the lower north-west corner of the Grand Gallery,—that the Great Pyramid had been entered with violence long before the time of Al Mamoon; and by men much better acquainted with the interior devices of Pyramid building, than were

[1] Murray's *Handbook to Egypt* (1867), p. 168.

that early Khaliph's enthusiastic Mussulman followers.[1]

After trying, testing, and being obliged to condemn, therefore, whatever can be said in favour of this new and forcible bringing up of a selected and long-after-date Arab supposed authority,—we find ourselves standing once again just where Professor Greaves stood in 1637 A.D.; or, obliged to reject every rag of testimony from the followers of a false prophet, as well as the classic accounts of Pagan Greek authors; and compelled to trust to modern observation, measurement, and induction alone, for all effective Great Pyramid research and explication.

The Coffer's Material.

'Nothing more than an old and dilapidated stone coffin,' said Sir J. Y. Simpson, when he displayed his perspective view of the coffer,—at the same time looking his very words.

The words are, however, ill chosen.

'Dilapidated stone coffin,'—implies something decaying away, possibly built up once of several

[1] *Life and Work at the Great Pyramid*, by C. Piazzi Smyth, vol. i. Plate 5; and vol. ii. p. 70.

pieces of stone, but now falling apart, and the material weak and perishable.

Yet the Great Pyramid's coffer,—at present an open, lidless, box-shaped, long rectangular vessel, —is cut out in a variety of red granite having the hardness almost of a gem. It is a very successful mechanical performance too, in the largeness, near regularity, and minute finish of the hollow, as compared with the comparative thinness of the sides,—seeing that it is all sawn or carved, without flaw or crack, out of one large block of the said adamantine material.

It still rings like a bell, of very pure and peculiar tone, when struck, as it is, unhappily, by almost every traveller who enters the room, on purpose to bring out that bell-like note. And finally, though it has been chipped, and considerably broken away even, at one corner,—though only as a desperately hard material, little chip by chip, through years of chipping administered by modern men with hammers,—that does not entitle it to the application 'dilapidated;'—for every particle of it, where touched only by nature, is still so hale, hearty, and sound,—that the whole structure would evidently have come down to us perfect, from its original artists—except for the mischievous pranks of multitudinous visitors during the last few centuries, and especially during the seventy past years.

One testimony, however, even these chips may give. For whereas Sir J. Simpson's picture showed the fractured portions of the coffer, exhibiting a dark and rotten-looking material inside, the outside being only a dull brown,—and thereby making the whole box rather remind one of a superficially somewhat whited sepulchre without, but which is full of black uncleanness within ;[1] —the ancient surface of the coffer is *really* a deep chocolate colour, and the new surface formed by the modern chippings, is comparatively of a light pink. So much lighter, certainly, than the outside, that in my photographs of the coffer by the magnesium light, where the ancient sides, exquisitely smooth from the remains of their former polishing, are dark and sombre in tone,—the inside material, as revealed by modern hammering, is positively brilliant ; totally reversing

[1] In the plan and elevation of the coffer, acting as the frontispiece of my vol. i. of *Life and Work*, and which plate had evidently been borrowed largely from for the above picture,—the broken parts of the coffer are generally represented very dark ; but only

therefore, one argument of the picture exhibited to the Royal Society of Edinburgh, by the advocate for depreciation, when he prayed the members so earnestly to pronounce for 'only an old coffin' verdict.

Tombic Hypothesis tried.

A 'mere coffin,' one is inclined to think, should not be very difficult to recognise anywhere; so much length, so much breadth, and so much height, are necessary to contain within, the full-grown human form laid out prostrate. Now the coffer in the King's Chamber in the Great Pyramid having about 78 inches of length internally, is fairly long; having 27 inches nearly of internal breadth, is fairly broad; but having something like 33 or 34 inches in depth,—is far deeper than there is any direct or known occasion for.

It is deeper too than any known Egyptian sarcophagus of nearly the same age: for the depth of that in the second Pyramid, measured from the top of its ledge to give it the utmost

on the principle of 'hill-shading' in maps; viz., to represent surfaces inclined in angle to the plane of the picture, and without any reference whatever to the physical nature of the surface.

favour, is 29 inches ;[1] of that formerly in the third Pyramid, 24·5 inches ; in the fourth Pyramid, 23·5 inches ; and in the fifth Pyramid, 25 inches.

Then there is something peculiar about the top of the coffer in the Great Pyramid ; a ledge is there found as if to receive a lid, fitting on, almost, but not exactly, like those of the sarcophagi of the second, third, fourth, and fifth Pyramids.[2] These sarcophagi all evidently have a coffin purpose, viz., to lock up, and keep safe, the mortal remains once enclosed ; and they accomplish this by a peculiar combination, first of dove-tailed grooves, which prevent the lid, after it is pushed into its place, from being lifted up ; and pegs arranged to fall from the bottom of the lid, into holes worked in the top of one side,—to prevent the said lid being dragged off sideways.

[1] See Howard Vyse's *Pyramids of Gizeh*, vol. ii.

[2] All these latter are abundantly described, and some of them pictured, in Vyse's and Perring's *folio* plates of the Pyramids. And I have tested, by the measures published in *Life and Work*, vol. ii. p. 272, the correctness of their large drawing of the sarcophagus of the second Pyramid. But the "delineation," and description of the coffer of the Great Pyramid by the above authors are mistaken. This so-called delineation, is merely a little figure introduced as a subsidiary feature into plans and sections of the King's Chamber ; and I should not have been hard upon it, or brought its littleness and superficiality into notice, unless a public attempt had recently been made to give it far more authority than it is entitled to.

But in the Great Pyramid coffer, though there are holes to receive such pegs, to prevent *horizontal* movement,—there are no provisions against *vertical* movement; for, the sides of the grooves being parallel with the sides of the coffer itself,—would offer no opposition to the lid being lifted straight up, if lid there were.

Now although no lid has ever been described to have been seen by any author whatever, to the coffer in the King's Chamber,—and although the French, in their great work on Egypt, engraved the said coffer as being in their day (1799 A.D.) without any ledge whatever,—yet we need have no difficulty in helping the tombic hypothesis to a lid; and then seeing what its friends will do with it.

On looking to the lids of sarcophagi in several other Pyramids, they are found to be 9 to 11 inches thick. But seeing that the Great Pyramid coffer has rather thinner walls than those vessels, we will make its lid thinner also; or, say 8 inches; and then allowing 1·7 inches of that, to be sunk into the ledge or groove cut out of the top of the sides,—there will be left outstanding, above the highest part of the walls or side of the coffer, a thickness of lid equal to 6·3 inches. And this thickness being added on to the outside height of the coffer, or 41·1 inches, gives 47·4 inches for the whole height of the coffer, when lidded or acting as a sarcophagus or coffin.

And then let us see what follows from that.

The entrance-passage of the Great Pyramid being only 47·2 inches high,—the coffer could not be got into the Pyramid at all as a sarcophagus, or with a lid duly fastened upon it so as to preserve the mortal remains inside safe from dust, accident, and vulgar gaze! The difference at the entrance-passage is certainly only 2-10ths of an inch, and *may* be charged on our estimate, for the thickness of a lid, not being true to so small a quantity;—but then, unfortunately, the door of the King's Chamber (now, by sinking of certain floor stones, nearly 43 inches high),—was anciently closer to 42 inches in height; or, too low by 5·4 inches of hard, undeniable, and unimpressible granite, to let the sarcophagized coffer pass into the chamber, where we now find it, unsarcophagized; *i.e.*, without a lid, and empty.

Into the King's Chamber, as a lidless box, the coffer might pass; but would have only the fraction of an inch to spare. And in a model which we have had prepared, of that massive but low doorway,—in the reality more than 100 inches thick of solid, well squared, and polished red granite,—when a coffer model is passed into it, the fit is so close, that it gives one strangely the impression, that the doorway was especially designed so as to admit the coffer vessel, uncovered by itself, or—lidless.

Hence, without presuming to say, that mortal remains of some one or other, either swathed in garments, or also packed in interior smaller cases, —were *not* once deposited in the smooth-surfaced, and inscriptionless *coffer* of the King's Chamber in the Great Pyramid;—yet we cannot close our eyes to the fact of there being not a little decidedly wanting in that vessel's character and claim to being perfectly and entirely suited for a *coffin only*. Or rather, some other and extraneous feature will always be cropping out, showing that something more than a coffin must have been intended, whether alone, or in conjunction with a coffin-function. But in either case, something much more noble, and vastly rarer to meet with, than a mere burial coffin.

Metrical Hypothesis.

If the tombic hypothesis, though, could not explain several important features about the coffer as we find it now, modern breakages excepted;—neither, it is affirmed by some, can the metrical idea; because,—'is there not a ledge cut near 'the top of the sides, as if for receiving, though 'imperfectly, a sarcophagus cover?'

True. But, on one hand, we have no proof that so small a portion cut out from the original box-sided vessel (and which cut-out might have been so easily inflicted afterwards),—was part of the original design; and it is not mentioned as existing, by any traveller until within the last forty years; while some of the older authorities distinctly declare against anything of the kind. And on the other hand, if *symbolisms* of weight and measure, are the chief objects of the Great Pyramid's metrical character, and not the furnishing to men actual working standards,—the partial ledge cut-out,—no matter whether a primeval, mediæval, or modern feature,—is of no serious metrical harm, for we can get at the intended idea of size notwithstanding.

To this end, very close linear measures are necessary. Such indeed have been made, and full particulars of them published in the second volume of *Life and Work at the Great Pyramid*, and they do indicate some most interesting metrical ideas. But since then, a learned Baronet has raised such a storm of ridicule against the said measures—that the author has been left standing almost alone, by the side of his long and laborious work.

'Though the coffer of the Great Pyramid was thus imagined,' said Sir J. Y. Simpson to the Royal Society of Edinburgh, 'to be a great standard of measure, Professor Smyth had 'cited the measurements of it, made and published by twenty-'five different observers, several of whom had gone about the 'matter with great mathematical accuracy; but all these 'twenty-five, as he owned, varied from each other in their 'accounts of this standard, in every element of length, breadth, 'and depth, both inside and outside. Professor Smyth had 'measured it himself, as a twenty-sixth measurer, and again 'his measure varied from every other.'

This statement was received with roars of laughter, quite unusual in a scientific assembly met for the prosecution of science for its own sake. But then the picture of, first, a confusion worse confounded before Professor Smyth went out to make improved observations; and then, as the result of his labours, that the matter was made even more confused than before,—this picture, painted in eloquent phrases, and given forth with a profusely smiling countenance,—was too much for the risible faculties of human nature (not inquiring very nicely into the truth of the statement) to withstand.

Now, although Professor Smyth in his first book on the Pyramid,—before going thereto, and when he was striving to make the utmost of all published observations by other travellers,—did collect and exhibit the measures of twenty-five observers, extending from 1553 A.D. to 1837 A.D., and some of them were very inaccurate indeed; he did not represent them as all *equally* uncertain. On the contrary, he showed that while some of the earlier observers varied from each other by 10, 20, or even 60 inches, and a few of the modern ones by 4, 5, and more inches,—there were three observers, whose care and experience bespoke every one's respectful attention to their results; and whose measures, for the inside of the coffer, did in fact come so remarkably close to each other, or within a very small fraction of an inch, with one single exception,—

that most persons would have allowed, on ordinary principles, that Professor Smyth had, even so far, rescued the coffer-capacity question from extreme uncertainty, and reduced it within narrow limits. Say, within a quarter of an inch in each element, of length, breadth, and depth; with the exception, that the latter feature was stated by the French Academicians as being 3 inches greater than what Professor Greaves, and Colonel Howard Vyse, had represented it to be.

Now, 3 inches of difference amongst men measuring down to, and often agreeing to tenths of inches, was too large a quantity to be anything else than an absolute blunder, of one party or the other; though which it was, who had been so unfortunate,—no one in society ventured to say,—until Professor Smyth went out to Egypt, and practically found that it was the French *savants* who were in that dilemma.

The French Academicians' measures in 1799, reduced to English inches, for the depth of the coffer, gave 37·285; while Professor Greaves, in 1638, had given 34·32—and Colonel Howard Vyse (with the moderate accuracy, but full honesty characteristic of a soldier), merely going to the nearest half-inch, subsequently gave 34·5 inches—Professor Smyth's measures, repeated over many parts of the coffer, to detect inequalities, gave 34·34 British inches.[1]

The first part of Professor Smyth's measures at the coffer, then,—instead of bringing back all the confusion of many, even the worst, of the twenty-five observers, with their whole armfuls of erroneous inches,—had the sobering effect of throwing out the only one error that remained, greater than the fraction of an inch, after his previous discussion.

He did not, however, stop there ; but entered immediately after, on a far more careful and complete metrical examination of the coffer, than any

[1] The actual numbers as given in *Life and Work*, vol. ii. p. 123, are :—

INSIDE DEPTH OF COFFER, MEASURED FROM LEVEL OF ORIGINAL, OR HIGHEST, TOP OF N., E., AND S. SIDES.

Part of Length where observations were taken.	Part of Breadth where observations were taken.			Means at each part length.
	Near East side.	Near Middle.	Near West side.	
	Brit. inches.	Brit. inches.	Brit. inches.	Brit. inches.
0·6 Brit. inches s. of inner N. end,	34·30	34·28	34·26	34·28
3·0 Do. do. .	34·44	34·36	34·35	34·38
5·0 Do. do. .	34·42	34·41	34·28	34·37
10·0 Do. do. .	34·40	34·38	34·28	34·35
24·0 Do. do. .	34·36	34·38	34·26	34·33
Mean at each part of breadth, .	34·38	34·36	34·29	34·34
General mean of all lengths and breadths = 34·34 British inches.				

one had ever made before; and was enabled thereby to bring to light, and has since published, many subsidiary features of its shape and size, over and above mere length, breadth, and depth or height, measured once inside and once outside.

In most inquiries, the circumstance of having eliminated all large differences amongst former observers, and entered upon the measurement of microscopic quantities, which reveal hitherto unexpected residual features, or anomalies they may be, in the subject under examination,—is usually held to mark an advance in knowledge. Not so, however, in the view of Sir J. Y. Simpson, when that improvement has been effected on the Great Pyramid coffer; for the person by whom it was accomplished on this twenty-sixth occasion of measuring, and whose measurements do 'differ from those of every other measurer,' in that they go into refinements which the older observers never thought about,—is held up to public ridicule precisely for so differing from his predecessors; and is laughed at accordingly, with a few exceptions, by a learned society at a scientific meeting.

Yet that despised person's observations, being both the last made, and the completest yet published by any one,—we shall make no apology for using them, in order to ascertain whether there was scientific foundation for the ridicule, which Sir J. Y. Simpson next cast upon the coffer itself.

Capacity of the Coffer.

'Surely,' argued the versatile Baronet, 'a measure of capacity should be measurable, but this coffer of the Great Pyramid seemed immeasurable—at least, to twenty-six different observers.' And then giving the differences of some of these from others, but in cubic inches of computed contents of the whole interior, such differences were made to look very overpowering; especially in the case of the French measure, for, as their actually three-inch mistaken quantity of depth was not eliminated, there was introduced from that cause alone, on so large a vessel as the coffer, a contents error of some 6000 cubic inches.

The coffer was also abused, for having the small subsidiary features of figure, discovered by Professor Smyth; and it was declared therefore to have an interior, 'apparently unmeasurable in 'a correct way by mere linear measurement—the 'only measure yet attempted.' 'If it was an 'object of the slightest moment' (said the popular Vice-President of the Antiquarian Society of Scotland,—with far more expression of vindictive hate, than archæological love towards this remarkable work of primal times), 'perhaps liquid 'measurements would be more successful.'

How liquid measurements are to be applied to test precisely the ancient capacity of a vessel, which has been broken away by modern hammers to nearly a third of its height at one corner, and injured over most of its top,—was not explained to the meeting. Nor was anything said, as to who was to pay for having so much distilled water, as the coffer full, conveyed into the interior of the Great Pyramid, or what methods were to be adopted to weigh or measure the said amount of water, with something approaching to perfect exactitude, when it had been got there.

In short, linear measure, duly applied in conjunction with reason, is really the only at present available plan,—and we shall soon see, if it be not sufficient.

Deviations, as discovered by Professor Smyth, from a pure rectangular and rectilinear-sided mathematical figure,—these undoubtedly are in the coffer; but the very fact that they had escaped previous observers, and were only found out by him, on comparing the results of his many measures, show that they are merely slight disproportions on the whole. Being effected too, usually by the occurrence of curves of long radius,—their influences are never so uncertain, as those of sudden or lawless irregularities of figure; and can be practically eliminated by making many measures over many parts of the surface, and then taking a mean. For a capacity measure, too, extreme regularity of rectilinear figure is not of any moment; indeed, it is rather undesirable, lest it should lead to the vessel being improperly used as a standard of length also. Now the linear standard of the Great Pyramid having been already laid down in the Pyramid's base-side,—we derive an argument rather in favour of, than against, a capacity measure in the coffer, on finding that none of its three linear elements, either for inside or outside, fall anywhere near an even round number of Pyramid inches, in terms of Pyramid numbers. We are narrowed thereby to question the coffer solely, on its adaptation to capacity purposes.

Now this research having been already carried

through in vol. iii. of *Life and Work*, and both the modern fractures and the 'sarcophagus-ledge-cut-out' having been filled up by calculation,—there result for the linear internal proportions of the coffer (in Pyramid inches)—

Length 77·85, breadth 26·70, and depth 34·31; which numbers being multiplied together, give as our first approximation to the cubic contents of the coffer, in the light of a capacity-measuring vessel, 71,317 cubic Pyramid inches.

But who will answer for the ancient coffer, when thus interrogated in *modern* times, telling truly what were the original size-intentions of its founder?

The said ancient coffer will answer both for itself, and its maker.

The Coffer on its own Capacity.

While the tombic idea gave a reason for the exceeding length, and moderate breadth of the coffer,—it gave no account of wherefore its large depth; wherefore also the thinness of its sides as compared with those of much smaller sarcophagi; and wherefore the bottom, very nearly an inch thicker than the sides.

The metrical idea, on the other hand, while it would not necessarily demand *a priori* so exceeding a length and small a breadth,—is quite content to accept them from another source, provided it be allowed to furnish itself with sufficient *depth*, to realize therewith and altogether, a certain predetermined amount of total capacity. Given therefore the length and breadth as already stated,—and then, both the large depth and some of those other features also, which were unamenable to burial ideas, are at once explained by the metrical hypothesis.

Only a few weeks since, the ingenious Mr. Henry Perigal sent us a copy of a card bearing date 1865,—and stating as follows :—

'ORIGINAL DIMENSIONS OF THE COFFER IN THE GREAT 'PYRAMID.

	Length.		Breadth.		Depth.		Volume.	
'Interior,	= 78	×	27	×	35	=	73,710	$= \frac{1}{2}$
'Exterior,	= 90	×	39	×	42	=	147,420	

∴ Thickness of walls = 6, and of bottom = 7.

'Bottom,	= 90	×	39	×	7	=	24,570	$= \frac{1}{2}$
'Walls,	= 2(90 + 27)	×	35	×	6	=	49,140	
							73,710	

'The exterior volume being thus exactly double the interior 'capacity, a solid exactly filling the inner space might be 'formed of the walls and bottom, which are likewise in the 'same ratio (2:1) to each other.'

If the above numbers (which were derived by their author taking the mean of Greaves, Jomard, and Howard Vyse's measures, and then choosing the nearest whole number) are British inches,—they give a vastly too great cubic capacity for the coffer; but they are not such inches; nor are they any uniform unit of measure, for they assume a slightly different proportion to each other and to inches, in every successive element of coffer measure. Yet are they, notwithstanding, so remarkably near—that ordinary observation would not have distinguished between a coffer framed on such ideal measures and the actual coffer restored.

The ideal one, too, has the merit of fulfilling by the nearest simple numbers possible,—certain geometrical requirements, as of bottom to sides; and of bottom and sides together to interior contents; and of interior contents to external volume;—and it fulfils all these with absolute accuracy.

This result constitutes in so far, a *geometrical* foundation for a *capacity* measure; and in so much as the block to form the ideal coffer, would be rather larger than for the real one (supposing Mr. Perigal's numbers to be Pyramid inches)—the real coffer block might in a manner be formed by trimming down the ideal one, to such size, as another and a quite different consideration might have rendered necessary for the absolutè size of a capacity measure. The real coffer is, we know, smaller; but is it smaller in each of its several parts, in such rather compound and varying proportions,—that all the mutual geometric relations of the ideal coffer, are still preserved?

The coffer shows that it is so, within limits of the errors of observation, for the following are Professor's Smyth's results extracted out of *Life and Work*, where they had been printed, in part at least, under no expectation of such a test as this,[1] and expressed in Pyramid inches.

	Length.		Breadth.		Depth.		Volume.	
Coffer Interior, =	77·85	×	26·70	×	34·31	=	71,317	$=\frac{1}{2}$ nearly.
Coffer Exterior, =	89·62	×	38·61	×	41·13	=	142,319	

And here we come at once upon some almost intentional results in the residual deviations of

[1] A portion of the test was contained in a paper published by the late Mr. Jopling in the *Leisure Hour* for 1863, p. 330, in so far as he produced a vessel with such sides and bottom,—that the capacity of the interior, was half that of the exterior. But he made the bottom less thick, instead of more thick than the sides: and was so far wrong both in external length and height, and internal depth; that it is futile to compare the real coffer with his theoretical creation, though very ingenious in itself.

figure discovered by Professor Smyth,—for unless the measured linear quantities for the exterior, had been duly corrected for the, only recently discovered, concavity of some of the sides,—the final cube quantity would have been grossly in error.

A still neater result, too, follows with the bottom and sides. For Professor Smyth had remarked, when measuring them in the Pyramid, that they were *not* exactly 6 and 7 inches respectively, in thickness; but were on the whole less, and by varying quantities, in either case;—all which much disappointed him at the time, but he printed the numbers nevertheless; and now has appeared a *reason why* they are so small; for without those diminutions, unequal diminutions too, on the 6 and 7 inches respectively,—the required capacity relations would not have been obtained even approximately.

As the real coffer, is an affair of practice and human measure,—there is no absolute exactitude even in the last numbers for it yet obtained. To procure the truest results, therefore, we must try to take all independent observations, of each and every component part, into account. Therefore, besides getting the thickness of sides and bottom by differencing the numbers already given (as in Mr. Perigal's manner, and which is quite sufficient in his theoretical case)—we also refer to the direct measures of the thickness of those parts given in *Life and Work*, vol. ii. pp. 120, 121, and obtain finally, in Pyramid inches,—

Thickness of walls = 5·952, and of bottom = 6·866.

Employing these quantities next in his second formula, we have,—

$$\frac{\text{Bottom,} \quad = \quad 89{\cdot}62 \times 38{\cdot}61 \times 6{\cdot}866 = 23{,}758}{\text{Walls,} = 2(89{\cdot}62 + 26{\cdot}70) \times 34{\cdot}31 \times 5.952 = 47{,}508} = \frac{1}{2} \left\{ \begin{array}{c} \text{very nearly} \\ \text{indeed.} \end{array} \right.$$

Within such limits then, as have been measured to practically, in the present day—the geometrical proportions of a very remarkable theoretical approach to the shape of the coffer, *are* preserved, in the real coffer,—its ledge-cut-out having been duly filled up.[1] We may now, therefore, make use of those proportions, in combination with former measures, for multiplying approximations to the *capacity* of the interior.

[1] The space contained in the ledge cut out of the sides, amounts to above 1200 cubic inches. And the amount of space it would remove from the interior contents of the coffer, if allowed to remove its thickness over the whole top, amounts to more than 2380 cubic inches.

Thus,

			Cubic inches.
1.	Contents of interior of coffer by direct measure,		= 71,317
2.	Do.	by half the exterior volume, . .	= 71,160
3.	Do.	by sum of bottom and walls, . .	= 71,266

There are thus three determinations, by mostly separate measures; and we may perhaps employ even a *fourth*. That 'ledge-cut-out' on the top of the sides, which was for so long looked on merely as a thorn in the side of the metrical hypothesis,—has now been found to be a most useful testimony to the capacity question; for, if a frame be made of such a size as closely to fill the ledge, round all four sides,—that frame will be found capable of just going, longways and in vertical plane, inside the coffer and diagonally from one corner to the other of it. Moreover, when so placed, and therefore standing on the bottom of the coffer, such frame seems nicely to represent the original height of the sides; an important element to have testimony borne to, so extensively has the old top surface been chipped away. Wherefore taking—

	Pyramid inch.
Mean breadth of cut-out on Eastern side from the several observations given in vol. ii. p. 116, of *Life and Work*, ...	= 1·63
Internal breadth of coffer, ...	= 26·70
Mean breadth of West side, as of sides already given,	= 5·952

the sum, or 34·282, gives us a new quantity for the depth; and substituting this in our first formula, we have $77·85 \times 26·70 \times 34·282 = 71{,}258$.

The mean of which 71,258, combined with the three preceding quantities,—yields 71,250 cubic Pyramid inches,—for our best determination by all Professor Smyth's observations, of the coffer's cubical interior contents. Each result, it will be observed, is perfectly certain about the 71,000, and the uncertainties in the three smaller numbers, look so like mere errors of observation, and oscillate by such small proportionate amounts (as $\frac{1}{700}$th part, or less) about the mean—that the final result, or 71,250 cubic Pyramid inches, may be looked upon as very close to the real size now, with the ledge and modern breakages filled up, and the real, as well as intended, size of old.

But still comes the question, is that quantity **71,250** cubic Pyramid inches, a mere accident of an accident;—or has it any standing ground of importance, when compared with either nature or art, or the doctrines of mensuration in capacity and weight?

Why of that Capacity?

The metrical theory of the Great Pyramid puts forward, that a combined capacity and weight measure standard there, should remind, both of the bulk of the earth, and its specific gravity.

Wherefore, taking this latter quantity at 5·7 times the weight of water at 68° Fahrenheit; and taking also the cube of the 10^{-7th} part of the *whole* axis of the earth, or of 50 Pyramid inches, and dividing by 10, we have the 71,250 of such cubic inches at once,—

or thus: $\frac{50^3 \times 5{\cdot}7}{10} = 71{,}250.$

The 5·7 for the specific gravity of the mean material of the whole earth, is not a settled quantity yet by modern science; because the question is in fact, rather above her highest efforts to do more than merely make distant approximations to. But while the above numbers are a fair representation, on the whole, of the rather mutually conflicting numbers of the late Mr. Francis Baily, of the Astronomer-Royal, and Sir Henry James; they have also some support from the already investigated relation of the whole Great Pyramid with the whole earth; for the former's proportion of being the $10^{5\times 8}$ part of the latter,—would have been thrown entirely out, by employing any very different quantity from 5·7; or, as we should write it perhaps, 5·70,—to show that we feel sure of it to hundredths as well as to tenths,—if such an explanation of the coffer is to be admitted at all.[1]

This mode, then, is both in principle and numerical coincidence perfectly sufficient for explaining why a standard of capacity and weight in the Great Pyramid should have its internal contents = 71,250 cubic Pyramid inches. But is there any opposing theory in the field?

Yes; $\sqrt[3]{71{,}250} = 41{\cdot}46$; and $\frac{41{\cdot}46}{2} = 20{\cdot}73$, or the length, very nearly, of the profane Egyptian cubit. Wherefore, say some, the coffer of the King's Chamber was made of just such a size that the internal contents should be equal to the cube of a length of two profane cubits. And of course they add, 'Therefore there is nothing scientific 'about it; metrical, it may be, in principle, if

[1] 5·69 for the earth's mean density, would give in the formula already detailed for the size of the coffer, 71,125; and 5·71 = 71,375; or outside all our measures, which vary from 71,160 to 71,317.

'you like, so as to contain 8 cubic profane cubits, 'but not of any high or noble scientific order.'

Here then we have once more, as with the *shape* of the Pyramid in the earlier part of our essay, two rival theories; each bringing out apparently the same 71,250 numbers given by observation. How then shall we determine which of the two ideas, the mind of the ancient designer was intent on memorializing?

Why, by looking round upon the accompaniments, and seeing if there are any *independent* hints to be derived from them.

It has been said for the above profane cubit hypothesis,—that the whole King's Chamber was laid out, so as to be exactly 20 such cubits long and 10 broad; showing also that the designer knew of such cubits, and employed them in his work.

Doubtless he did so; but yet had no intention of raising them to the honour of deciding the contents of the coffer. Modern measure shows[1] that the breadth of the chamber, is to 0·1 of an inch, exactly half of the length; and that the one therefore represents 10 of precisely the same

[1] *Life and Work*, by C. Piazzi Smyth, vol. ii. pp. 101 and 340.

linear units or standards that the other contains 20. And being thus repeated twice with such perfection, that standard may be considered well established. What then, by all means, is the length of that standard in Pyramid inches?

Reference to *Life and Work*, vol. ii., shows that it amounts to 20·61 Pyramid inches; and $20·61 \times 2$ and raised to the cube power $= 70,036$; or a quantity totally insufficient to stand for the 71,250 of the actual coffer.

The profane cubit, then, of that very chamber, failing sadly,—let us look for further hints.

1. Over the entrance-door of the King's Chamber, outside, are engraved four deep vertical lines, dividing the space into five equal parts; and forming therefore, apparently, a sign to those who enter that chamber, that they shall find a reigning principle of 'a division into 5,' pervading it.

2. The walls of the room, all round and round, are divided into 5 equal horizontal courses; again reminding of a division into 5.

3. From the chamber's lowest course, which is 47 inches high, a depth equal to 5 inches is taken off, by the mode of introducing the floor; and this amount again reminds both of 5, and of

Pyramid inches; for these units, instead of being at 6's and 7's, go into the earth's axis length with 5's and 10's.

4. The said 5 inches being taken off that lower course of the chamber walls,—the remainder course forms a sort of large tank enclosing the coffer; and is equal, in cubic contents, to 50 times the interior bulk of the coffer.[1]

5. The method of construction of the whole solid substance of the Great Pyramid, is in vast layers of horizontal courses, extending throughout the entire building; and of those courses (if the mean of the best measures yet made can be trusted) 50 intervene between the foundation plane of the Pyramid, and the floor of the King's Chamber carrying the coffer.

There, then, may still be seen most striking references, undoubtedly too by the original builders, to divisions either by 5 or 50; and they are repeated five times over. And while no such division was known to the profane

[1] See *Life and Work* for the particulars, vol. iii. p. 168. Here, we may merely state that there are two ways of taking the measurements, of which one method gives $\frac{412{\cdot}14 \times 206{\cdot}09 \times 41{\cdot}9}{50} = 71{,}178$; and the other $\frac{412 \times 206 \times 42}{50} = 71{,}292$; mean, 71,235.

Egyptian cubit,—it is the very essence of the Great Pyramid's scientific system.[1] Concealed that system may ever have been from man, and even Egyptian Pagan men: yet surely enshrined in the very foundations of the building, and now coming forth to view, simple, grand, and duly attested, when questioned by number, weight, and measure.

The Latest Discovery.

Even when printing these pages, another numerical confirmation of more than 'an old coffin' object, has appeared. For whereas satisfaction might well be felt, at finding, as in the previous pages, that the ostensible irregularities of the coffer, had useful purposes to perform towards its final result, or were really important adjustments of size,—also that the commensurability of the room's lower course, by an appropriate number, with the coffer,—proves that the coffer is the intended vessel for that room (and intended too with the ledge filled up); and also that the similar

[1] The profane Egyptian cubit was divided into palms, and these again into digits or dactyls. Of these last there were four in one palm; but the number of palms in the cubit is given by some authors at six, and by some at seven. The origin it may be of confusion being at sixes and sevens.

number of the masonry course forming the pavement to the coffer level, shows that the construction of the whole body of the Great Pyramid to that height is both suitable to, and symbolical of, the coffer's theoretical origination,—the news now comes from Mr. St. John Day, that the coffer connects itself scientifically with the whole shape of the Great Pyramid, and that a shape which discriminates it from all other Pyramids.

The shape of the Great Pyramid, already set forth at p. 183 to 193, is characterized, as first published by the late John Taylor, by affording an indication of the mathematical quantity π. And now Mr. St. John Day finds the same proportion in a combination of the height, with two sides of base, of the coffer. Only very close when he uses the *maximum* figures of the several lengths, breadths, and heights given in vol. ii. of *Life and Work*; but precisely proving thereby, as he considers,—that it was *necessary* for the primeval architect to construct the coffer with the apparent irregularities now found, in order that in one and the same vessel there might be,—

1*st*, A definite amount of cubical contents, settled by an earth-reference formula.

2*d*, A certain inter-commensurability of parts, to prove the contents, and disprove the effects of modern mischief. And

3*d*, A sign to show that the coffer we now find in the Great Pyramid, is the original one ; viz., the right coffer, in the right chamber, in the right Pyramid, and suitable to no other place under the sun in an equal degree. And

4*th*, Perhaps a length, breadth, and extra depth, permitting of a possible honourable burial to one who had worked nobly ; and called attention to himself by no vain-glorious inscription.

As Mr. St. John Day is shortly to publish his discovery with the Philosophical Society of Glasgow, it would be inexpedient to say more now.

We conclude this section therefore with restating :—1*st*, that the coffer's interior volume, derived from $\frac{50^3 \text{ Pyramid inches} \times 5{\cdot}7}{10}$ represents the intended cubic size or true contents of the Great Pyramid standard of *capacity measure.*

And that, 2*d*, the weight of that volume of water at 68° Fahrenheit, and at the mean Pyramid barometric pressure, will be the weight of the Great Pyramid standard of *weight-measure.*

Therefore $\frac{71,250}{5 \cdot 7}$, or 12,500, will be a number of cubic Pyramid inches of matter, equal in density to the mean density or specific gravity of the whole bulk of the earth;—and which 12,500 cubic inches will weigh as much as the coffer's contents of water, at the said temperature and pressure.

If, too, we further assume, that such grand standard of Pyramid weight-measure, shall be divisible into 2500 parts; and each of these parts be denominated a pound weight of the Pyramid,—both the Pyramid system of numbers is fully preserved, and such Pyramid pound will not only measure, within $\frac{1}{30}$th part, the same as an avoirdupois pound (either by the latter's happy accident of birth, or perhaps through the care of traditional preservation);[1] but will be scientifically definable, amongst nations in a high state of civilization, as, 'the weight of *five cubic Pyramid* '*inches* of the *earth's mean density*;' the latter indicating rock-material of approximately twice the specific gravity of basalt, taken at the mean temperature of the surface of the whole earth.

[1] The late John Taylor's discovery of the near commensurability of the British *Quarter* for corn-measure, with the *fourth* part of the coffer of the Great Pyramid, should also be borne in mind, as it still holds good.

INTELLECTUALITY OF GREAT PYRAMID—*continued.*

CHAPTER 30.

UNITS AND STANDARDS OF GREAT PYRAMID MEASURE—*continued.*

Heat.

THE Pyramid deduced standard of heat unites in itself the following recommendations :—

1. A truly Great Pyramid division of a natural scale ; viz., one-*fifth* the distance between freezing and boiling of water, above the former.

2. The mean annual air-temperature surrounding the Great Pyramid.

3. The mean annual air-temperature of the whole earth, taken according to its habitable surface.

4. The whole scale from freezing to boiling being imagined to be numbered pyramidally in even degrees from 0 to 250°, the standard point

reads 50° on that scale, and the point of greatest density of water reads 10°. All the notable referring points are thus marked in properly Pyramidal, and symbolical, numbers of the Pyramid; unlike the French centigrade and decimal division, which has the greatest density of water at 4° of that scale.

5. The standard heat-reference at the Great Pyramid, being its own 50° (68 Fah.), is a peculiarly suitable temperature for men to work well in, either with mind or body.

But the French metrical system, of making their heat-reference for some things the greatest density of water (39° Fah. nearly); for others the freezing of water; and for others again some anomalous temperature (between 40° and 50° Fah.) —where their erroneous metal metre becomes by expansion approximately correct—this system invites man to observe under numbing and untoward influences, contributing thereby needlessly to his pain, perplexity, and trouble.

To compare small things with very great, or with religious: the heat part of the French metrical system, acts towards men, much as does the Hindoo religion towards its devotees. That religion requires its holiest followers to torture themselves violently, to sleep on spiked beds, to stand with one arm oustretched until it has become rigid; and to swing on poles suspended from hooks entering their flesh; and what sort of religious notions does it give them after all! But the Christian religion draws by love, gentleness, cheerfulness, charity; it induces its devotees to walk in the ways which lead to happiness; and to make all the good, true, and pleasant things of the world clean and allowable unto them. So likewise, *mutatis mutandis*, the temperature arrangements of Great Pyramid metrology.

In the middle ages, when science did *not* prosper, and when astronomy waned, but astrology flourished,—no man's wisdom was thought anything of, unless he put on large spectacles, an imposing dress, and taught in a dead language which the vulgar could not comprehend. And so, a scientific observation in physics, can hardly be truly conducted at the French centigrade standard therefor, without a man burying himself learnedly in an ice-cooled chamber, far from the amenities of ordinary life.

But the Great Pyramid system of heat-reference rather encourages each would-be observer, to seek out,—or remain in, if he has already met with,—the most comfortable state of temperature possible. Whenever he finds himself so situated, whether in a regular observatory, or his private house,—he is also under suitable circumstances for observing agreeably with the Pyramid temperature standard. And when the circumstances appropriate for observing occur thus daily and without special effort—they are likely to be used more frequently ;—even as science is much more discussed now, when it is spoken of and prosecuted in our mother-tongue,—than ever it was, when confined of old to a few educated men, speaking stiffly in Latin or Greek.

INTELLECTUALITY OF THE GREAT PYRAMID—*continued.*

CHAPTER 31.

UNITS AND STANDARDS OF GREAT PYRAMID MEASURE—*continued.*

Angle.

ONE of the least exceptionable parts of the French metrical system, was probably the strenuous attempt made to drive out of use the Babylonish sexagesimal degrees, minutes, and seconds for angle—by introducing a so-called decimal graduation of the circle. In this graduation, the degrees were indeed subdivided decimally, but there were 400 degrees in the whole circumference. Hence perhaps, in part, the idea failed to be accepted even in France.

The Great Pyramid system, on the contrary, regards the circle circumference as its grandest unit of angle, and divides it into 1000 degrees accordingly. To these we may add decimal subdivisions,—and there result thereupon, both expressions for the chief Pyramid angles, to less than the nearest tenth of one of those small degrees,—though the same angles are excessively rough when given in terms of any other angle-graduation system yet tried by man;—and there are also some facilitations for navigation and itinerary measures, in the very near commensurability of what may be called Pyramid miles and Pyramid knots.[1]

[1] See *Life and Work at the Great Pyramid*, vol. iii. pp. 202-215.

INTELLECTUALITY OF THE GREAT PYRAMID—*continued.*

CHAPTER 32.

UNITS AND STANDARDS OF GREAT PYRAMID MEASURE—*continued.*

Chronology.

THIRTY years ago, Sir John Herschel called attention to the entrance-passage of the Great Pyramid, as indicating on astronomical grounds the date of the building's foundation ; and if that be fully borne out,—what a boon both to Egyptology, and indeed to all the early history of intellectual mankind ; for that is confessedly an inquiry where, in the absence of records of eclipses, or any other positive method, even the best authors are known to take in, or drop out, of their reckonings, whole thousands, even many thousands, of years ; and they do it with a degree of freedom which as much astonishes the unlearned public,—as it utterly confounds their rivals in the same branches of research.

Sir John Herschel's argument seems to have been,—so far as we can make out, though it is exceedingly desirable for general information that that eminent *savant* should re-state his views—

1. The Great Pyramid is to a certain extent unmistakably astronomical, from the near approach which the sides of its base make to perfect cardinal orientation.

2. The entrance-passage is approximately in the plane of the astronomical meridian, and looks in that plane, very definitely to a certain small angular distance underneath the polar point of the sky; or suitably to a transit of the meridian beneath the pole, by a circumpolar star at a given polar distance.

3. At a certain date B.C., and which Sir John Herschel, in 1839, considered was the most probable, out of several then published literary dates, for being that of the foundation of the Great Pyramid,—he discovered by calculation, that a very remarkable star, *a* Draconis, was at precisely the angular distance from the pole of the sky, indicated by the entrance-passage.

First Order of Objections, and their Answers.

With some men of astronomical knowledge and mathematical ability, the above considerations seem to have been held satisfactory.

But with others, who are great in antiquities of the nick-nackitarian kind, the practice of medicine, and divers pursuits, useful enough, but of inexact foundation,—a directly contrary effect appears to have been exerted; nor, on several recent occasions, could too much contempt be shown by these gentlemen for the commission of, in their eyes, such an Egyptological solecism.

'Why,' have exclaimed members of this party, even up to a few weeks ago, 'the slope of the entrance-passage of a 'Pyramid was intended for nothing but lowering a sarco- 'phagus down to its final rest by! What an absurdity to be 'inventing astronomical explanations of such simple mecha- 'nical matters! Besides, the Egyptian chronology, from one 'end of the nation's life to the other, was attached to one only 'star in the sky, and that star was never seen from Northern 'entrance-passages, for it was the Southern Sirius, or Sothis.'

Yet before going into our chief astronomical proofs for Northern *α* Draconis,—let us try to please these gentlemen, by testing their sarcophagus-lowering idea of the Great Pyramid's entrance-passage, on truly Egyptological grounds; as thus —

1. Said entrance-passage was so low in the roof, that the coffer of the King's Chamber could not have entered it, when in coffin guise, or with an ordinary sarcophagus lid fastened upon it. (See p. 283.)

2. In neighbouring tombs, sarcophagi have been found in such situations, that they must have been moved to them partly by descending vertical shafts, and partly by advancing along horizontal galleries. Whence it would appear that a special and peculiar sloping path, like the entrance-passage of a Pyramid, was not absolutely necessary for moving a sarcophagus.

3. Even granting that a sarcophagus was taken piecemeal into the chamber, by the too narrow passage; and the intended King's corpse introduced in the same manner, and packed into the big sarcophagus there in the chamber; also that that burial, and the safety and security of the embalmed body, were the whole objects of the Pyramid,—then, why was the entrance-passage left, or indeed ever made, a conspicuous and permanent feature of the Pyramid's architecture?

As Lord Valentia truly remarked at the beginning of the century—these Pyramid entrance-passages, lined with fine white stone, have served to lead plunderers right up to the very room, where, on the burial hypothesis, they ought not to be allowed to go on any account. While, had the passage merely been obliterated, after the deposition of the corpse, by being built up with ordinary masonry,—half the gigantic building, or more, would have had to be knocked down, before any clue would have been obtained as to the whereabouts of the corpse.

To the tombic hypothesis, therefore, as the *sole* object of the whole structure,—all the most striking, pervading, and positive characteristics of the entrance-passage of the Great Pyramid, (and its features were apparently copied, for the mere sake of copying, though without understanding, into other subsequent Pyramids,)—are a continual obstacle.

Not so, however, to the astronomical chronology department of the Great Pyramid's metrical theory; (and equally whether a little burying was performed inside some part of the building or not). For,—

1. This metrical theory utilizes at once the important fact, which the Egyptologists could not

deny, but tried to throw contempt upon,—viz., the position of the entrance-passage in the plane of the astronomical meridian—to within so near an approach to exactness as 5′ of azimuth angle. And,

2. The *permanence* and grandeur in design[1] of the passage's surrounding and formative architecture (though the passage's own bore might have been afterwards filled up and rendered impassable for ages by long plugs of stone thrust down it); and which distinguishing architecture is so important for memorializing to distant posterity the exact direction of a certain star, where it was once seen at a certain date, in a crucial astronomical position.

The metrical theory, starts thus with an eminent claim, in point of principle, to whatever is permanent or peculiar in the structure and position of the entrance-passage of the Great Pyramid having been built upon its account.

And next, Sir John Herschel's citation, as already given, of the *last*[2] lower meridian passage of *a* Draconis, as the circumpolar star of

[1] The floor of the passages is laid 400 inches broad, though only 41 inches thereof are visibly employed; see *Life and Work at the Great Pyramid*, by C. Piazzi Smyth, vol. ii. Plate 2, and pp. 11-43.

[2] See *Life and Work at the Great Pyramid*, vol. iii. pp. 260-285.

the Pyramid-building day ;—has been notably strengthened *Egyptologically*, since he wrote, by the publication of William Osburn's *Monumental History of Egypt*. For therein, and quite independently, that thorough hieroglyphic scholar, after a searching examination of all the authorities, places the fourth, or Great Pyramid building, dynasty in the very period from 2200 to 2100 B.C., when astronomers compute that the star *α* Draconis actually had the particular polar distance which is indicated by the entrance-passage, combined with the latitude of the place.

Further Objections and New Answers.

But the Egyptologists of the French and German schools still treat Sir John Herschel and *α* Draconis with disdain, while they go on asserting for the date of the Great Pyramid's building, not 2170 B.C., but so much as 3400 B.C., 4500 B.C., or almost any other large number of years. They sometimes too wind up their objections by asserting, that the peculiar angle of elevation of the entrance-passage is architectural only, or probably accidental, rather than scientific; and that, as *α* Draconis has been in successive ages during the last 5000 years at all distances from the pole, between one degree or less, and twenty degrees,—of course it must have been at some time, coincident with the direction of the entrance-passage, which lies somewhere or other between those broad limits.

This might have been a pretty good argument at one time; and when excluding every other consideration than merely the polar distance of *a* Draconis and the angle of the entrance-passage; because evidently, within certain limits, no matter to what angle and direction, in the meridian plane, the said northward-looking passage lay,—*a* Draconis would have been seen coincidently with it at *some* chronological period or other.

But now, thanks to recent researches,[1] we may turn upon the Egyptologists and ask them,—If the angle of elevation of the entrance-passage was an entirely accidental, or at least a perfectly unastronomical matter (although in the plane of the astronomical meridian), and might have been anything else, as easily as what it was made to be,—how did it come to pass that of all the possible angles, between 1 degree and 60 degrees,

[1] See *Life and Work at the Great Pyramid*, vol. iii. pp. 260-285, by C. Piazzi Smyth.

such a particular, most unique, and all-important angle as 26° 18′ was hit upon, so as to chime in precisely with, and call attention to, several other simultaneous astronomical phenomena and arguments. For in the year when *α* Draconis on the meridian but beneath the pole, was last seen at precisely that passage elevation angle of 26° 18′ —such polar phenomenon had then, and only then, the important complement for time-observation, that the Pleiades were simultaneously crossing the meridian above the pole and near the equator. And, what is even more important still for chronological purposes, the Pleiades were then,—in that one year alone, of all years for more than 10,000, past and to come,—in the meridian of the equinoctial point; or this said equinoctial point, the beginning of all reckoning by right ascension in the sky, was on the meridian coincidently with the other two then also mutually agreeing stellar facts.

Not only therefore was a triple time and space phenomenon marked, by the one selected angle of the passage; a celestial combination, too, perfectly capable of fixing a date beyond all question, or all repetition of the numbers of the cycle through human history, so far as such history has yet been recorded,—but a chronological method for all years is indicated, in the annual increase of distance of the Pleiades group from the equinoctial point, at the rate of about 3·5 seconds of right ascension every year. The Pleiades, in fact,[1]—with the assistance of the precession of the equinox which causes them to perform that apparent cyclical motion in the sky with the period of 25,860 $\pm x$ years,—form the Great Pyramid clock; and that clock was beginning its mighty round, or had its hands at 0h. 0m. 0s. when α Draconis was last at the distance from the pole marked by the Pyramid's entrance-passage; or, as we will follow Sir John Herschel, and as he implied on far less amount of data,—'when the Pyramid was being built.'

Still to persist in saying that all this took place by accident,—and that the Great Pyramid was founded thousands of years earlier than the date (2170 B.C.) thus symbolized by permanent built lines in the Pyramid, and computable by modern astronomy—is what we should hope that no one capable of making induction from obser-

[1] See R. S. Haliburton, on the 'Year of the Pleiades,' in *Life and Work at the Great Pyramid*, vol. ii. pp. 370-448.

vations in exact science, will be found wilfully doing;—especially as we may add, that at the rate of a Pyramid inch, for a year,—the number of years in the whole precessional cycle is found built into the sum of the two diagonals of the base of the great monument. Also that the Grand Gallery, the most remarkable of all the passages of the interior of the Pyramid, and leading away southward, in the meridian plane, from the chief junction of the passages,—is by many considered, with the seven overlappings of its long and solemn walls, to be a reminder of the Pleiades seven stars, so famous in Eastern tradition from the days of the patriarch Job; while the north end of the Grand Gallery, subtends from the centre of the base of the Pyramid, the upper and lower culmination of *a* Draconis at that particular epoch of temporary time-agreement with the Pleiades.

Summation for a few of our Readers.

This then is the continually rising manner, in which the Great Pyramid, when tested against the severities of modern science, comes forth successfully, and forms a system peculiarly its own; resting on no other basis than the measurable mathematics and physics of its structure, combined with the great cosmical facts of nature.

Hieroglyphics and Egyptology have no place in the affair. Those sounding literary studies have claimed the Great Pyramid for ages—but could never make anything out of either its inscriptionless walls or peculiar proportions. And neither ancient Egyptians knew, nor do modern European literary scholars nor even the present inhabitants of Egypt know, anything certainly about the origin, meaning, or the exact nature and numerical measures of these things.

The scientific symbolizations of the Great Pyramid, were not only at all times entirely hidden in profane and idolatrous Egypt,—but breathe a perfectly different spirit, from what all the ancient Egyptians either professed or practised.

Let no one therefore waste his time, as did a foreign contributor to a recent number of the Proceedings of the Royal Society of Edinburgh,—by suddenly finding out that the Great Pyramid's deduced, symbolical, scientific, and earth-commensurable cubit, was not the common agricultural cubit in use throughout historic ancient Egypt!

Of course it was not: and that is the principle, nay the virtue, of the whole of the Great Pyramid metrological arrangements, founded on earth and heaven commensurabilities,—viz., that none of them could ever have been discovered by, or were ever known to, any of Egypt's idolatrous sons. While a great and now accumulated strength of the scheme or theory is, that one and the same principle runs through every portion of it.

Thus of the five major features, already discussed, as the surest facts pervading the whole Great Pyramid,—

1. The very early Egyptians marked nowhere their knowledge of π; or appreciation for *Pyramid numbers* in multiplication and division.

2. Nor their knowledge of the sun's mean distance.

3. Nor their appreciation of latitude with astronomical orientation.

4. Nor their knowledge of the earth's weight.

5. Nor their acquaintance with the earth's mean temperature.

And of the five minor,—

1. They did not use Pyramid inches, or Pyramid cubits of 10^7 to the earth's polar semi-axis:

nor reckoned how many of them were meted out by the earth in a day's mean motion round the sun.

2. They did not use Pyramid capacity and weight measures: nor arranged their own in divisions of 5 and 10.

3. They did not mark anywhere their attention to a special standard of temperature, and that in a Pyramid division of the water scale of heat.

4. They showed no method of graduating the circle; much less in Pyramid numbers.

5. Their one star of observation was neither *a* Draconis nor the Pleiades, but Sothis the Dog-star; nor was their great cycle of time the precessional period; but the wretchedly short Sothiac circle of 1461 years.

In not one of these Great Pyramid points is the wisdom or folly of profane and idolatrous Egypt foreshadowed at the Great Pyramid,—but something infinitely better and higher, as well as radically distinct in every respect, appears. Appears too, out of actual modern measured, and still measurable, facts of that most ancient building, the Great Pyramid, when compared, as already stated, with modern science, and her now advanced knowledge of nature. These builded facts too of the Great Pyramid, while they are facts, have had no other, than the metrical, explanation ever given for them; they are useless or objectless to the tombic theory, and yet they are there. No one can deny that those ancient metrical facts of form and size are there, though many do try to obscure their presence or conceal their nature, by throwing them into averages of many other, and very different, Pyramids to the Great one. Even this refuge, however, is now beginning to fail; and recourse is had to asserting, that 'the 'deception of the case consists in the facility 'with which modern science can find out some 'resemblances to some of its formulæ almost 'anywhere.'[1]

When any other ancient building shall have been brought forward, and had as much openly proved for it, on one grand, consistent, and connected theory, teeming with a similar *consilience of inductions*, and within the same small limits of numerical error, as we here have been enabled to set forward at length for the Great Pyramid—the

[1] See Appendix 3, for a powerful paper on this subject by W. Petrie.

above objection may be considered argument: at present we fear that it flavours more of prejudice.

But can we assist its authors, who have not yet produced a practical example proving their side of the question, by searching them out something whereupon they may test their assertions?

Conscientiously we will try.

CHAPTER 33.

INTELLECTUAL CONTRAST.

WE are now to look forth through all the world, for any other ancient building which may compete in intellectual majesty and scientific character, assignable from modern measure and discussion, with the Great Pyramid of Jeezeh.

If such there exists, even within a couple of thousands of years of the date of that building, far be it from us, to throw any veil over its excellencies. But in what land is it to be found; for have we not already passed all other countries in review, both architectural and archæological, without making any such discovery? A titled gentleman has, indeed, lately asserted before the Royal Society of Edinburgh,[1] with reference to the mere chambered tumuli on the banks of the Boyne, and which rounded heaps of loose and soft material,

[1] On December 16, 1867. See Appendix 2.

he spoke of as 'Pyramids,'[1] that 'if Professor 'Smyth would visit and measure *them*, he would, 'perhaps find them as interesting and astonishing 'as others he had visited, or even more so,' *i.e.*, than the Great Pyramid of Jeezeh. And again, on another occasion,[2] the same popular chief of the medical profession declared before the same Society, —'that at the time at which the Great Pyramid 'was built, probably 4000 years ago, mining, 'architecture, astronomy, etc. etc., were so ad-'vanced in various parts of the East as to present 'no obstacle in the way of the erection of such a 'noble royal mausoleum as the Great Pyramid 'undoubtedly is.'

Respecting the first of these assertions, Professor Smyth can only look upon the idea of applying accurate measure to mouldering heaps of soft matter without any fiducial referring points, or any fixed external characteristics,—as a mockery; and he cannot think of depriving Sir J. Y. Simpson of the responsibility of proving practically to the world, that he meant it in a better sense. While, as regards the second assertion, it

[1] And has since insisted, in the Proceedings of the same Society, on the propriety of the term.

[2] January 20, 1868—conclusion of address in Appendix 2.

has simply been asserted by the eminent Baronet, and he is bound, as a true and honest man, to prove it also; calling the Great Pyramid, if he will, 'a royal mausoleum,' but not omitting to show his promised various Eastern habitats of, and abundant contemporary parallel examples for, all the astronomy and physics actually found to reside in the Great Pyramid's measured proportions; partly illustrated, as they are, by the five major and five minor characteristics of surpassing intellectuality, pointed out in our chapters 22 to 32.

Meanwhile, we can only proceed with the nearest building in point of time known to ourselves, or generally considered at all worthy of being compared against the ancient wonder standing in the land of Egypt; and we commence a comparison accordingly in a plain manner, but to the best of our knowledge of the facts.

Babylonish Example.

The example we allude to is the Planet-temple of Nebo at Borsippa; already touched upon in chapter 10, but deserving of some more notice, from its high appreciation in several modern literary quarters.

There is the disadvantage, no doubt, for simultaneous comparison, that the Borsippa temple is 1700 years later than the Great Pyramid; but such circumstance must be so manifestly to the prejudice of the elder, and advantage of the younger, building,—if, indeed, it be true, as some popular philosophers protest, that all scientific knowledge ever found amongst men, is an affair of their own progressive development and creation of it,—that any involuntary partialities of our own for the Great Pyramid, must be more than neutralized.

Is, however, the Borsippa building the best example of that later age during which it flourished, in the estimation of its own friends and supporters?

What better answer can be returned, touching the architect, at least, of Borsippa, viz., King Nebuchadnezzar,—than the Bampton lecturer's recent eulogy of him in the city of Oxford; an eulogy too, touching that very qualification of *building*; or in these words,—

'Nebuchadnezzar, a sovereign who, to the military talents 'of an able general, added a grandeur of artistic conception 'and a skill in construction, which place him on a par with 'the greatest builders of antiquity.'

This high praise, after so many ages of oblivion, would have been sweet indeed in contemplation, to that awfully vain autocrat; of whom the Scriptures relate (and there in contemporary record),—that when he was walking in the palace of the kingdom of Babylon, the King spake and said,—

'Is not this great Babylon, that I have built for the house 'of the kingdom, by the might of my power, and for the 'honour of my majesty?'

Or, to use the identical Nebuchadnezzarian words, as recently read off from cuneiform inscriptions,[1]—

'In Babylon, the city which is the delight of my eyes, and 'which I (Nebuchadnezzar) have glorified. A strong fort in 'brick and mortar, in strength I constructed. Inside the 'brick fortification, another great fortification, of long stones, 'of the size of great mountains I made. And this building I 'raised for a wonder.'

This was what the king said, of, and for, himself. But what does modern science say, touching the head and crown of all this bombast,—viz., the Planet-temple of Nebo at Borsippa?

1*st.* Modern science, we suspect, is a little at variance here with some scientific men. For

[1] Chiefly by the admirable and untiring labours of Sir Henry Rawlinson.

while those in London, have hitherto praised the Borsippa temple exceedingly, for, 'the practical 'success and *nearness* with which it was placed 'in the astronomical orientation desired,'—though they knew it was more than four degrees out, and agreed to consider that a small error;—the Great Pyramid, on the other hand, has either been passively neglected, or actively abused in other circles, 'for its large and utterly vitiating errors in orientation,'—although they were equally well known to the critics to be under five minutes of a degree.

Yet the fact remains, that the error on the sort of astronomical orientation intended at the Great Pyramid was sixty times smaller, than was committed in a somewhat similar case, 1700 years later in the world's history, by the most powerful king, learned priests, and wealthy community then existing; and the better workmen will have justice done to them some day, if not now.

2*d*. Even if the Babylonians had hit the sort of orientation they aimed at, much more precisely, —that position was, in its diagonal character, of a most obstructive nature to any observation of, and for, what M. le Verrier calls 'astronomy of precision;' that is, the astronomy which by observations in the direction of the cardinal points, and chiefly in the plane of the meridian, ascertains most accurately the absolute places in the sky, of the stars and other heavenly bodies, at a given epoch; thereby furnishing the only sure foundation for, either correct chronology through the longest intervals, or safe theorizing on all the grander secular changes in both the solar system and the starry heavens.

3*d*. The Babylonian building, even apart from its outward orientation, had no truly scientific character about its general construction. Its surface was lop-sided, and broken with corners of terraces, pannelled walls, dwellings for priests, and flights of stairs,[1]—so that not even its warmest admirers have pretended to say there was anything scientific in the shape, though that is so mighty a feature for proving high intellectuality in the conception of the Great Pyramid.

But they do say for the Borsippa building, that, being dedicated to the seven planets of

[1] See Mr. Fergusson's restoration at p. 139, vol. i. of his *Architectural History*.

Chaldæan astronomy, each stage was appropriately *coloured.*

Appropriately coloured! a surface demonstration only; but of what nature and with what object?

Why, the lowest stage, we are told, was coloured black, as symbolical for Saturn; the next orange, for Jupiter; the third red, for Mars; the fourth yellow, for the Sun; the fifth green, for Venus; the sixth blue, for Mercury; and the seventh white, for the Moon.

A very ill-assorted list of so-called planets surely, and extremely deficient without Uranus and Neptune and the Planetoids,—especially for a people supposed, by their friends, to have discovered the vastly fainter 6th and 7th satellites of Saturn. But, accepting the planets as given above in their diverse-coloured stages, what learning were these colours intended to typify?

And then comes the sorry confession, that each hue was supposed to be symbolical of such planet's influence on human well-being, both moral and physical; and therefore was superstitious, astrological, and nonsensical, rather than astronomical.

4th. Though some mere chronicling of eclipses

and extra-meridian star-gazing went on from the upper tower, as an occasional occupation of its inhabitants, rather than a standing feature of the office itself,—there was much more of necromancy, idolatry, and immorality.

5th. Though the adventitious claim may be put in,—rather for the city generally, than for the temple of Borsippa in particular (and certainly not as provable by any measures yet made upon any Babylonian temple's walls or mass),—that a general metrological and numerical system was invented and established there,—yet that system was such, that modern science cannot approve of it.

Its worldly success was great, no doubt. Commenced in Babylon in or about the neo-Chaldæan times, and remarkable chiefly for its sexagesimal arrangements and subdivisions, it replaced there an older decimal method; and so took the fancy of the Greeks, who followed Alexander the Great's army, that it was adopted by them, taken back with them to Greece, and allowed there to enslave the conquerors of its native land.[1]

[1] Dr. J. Brandis of Berlin (1866) on *Münz—Mass,—und Gewichtswesen in Vorderasien, bis auf Alexander den Grossen.*

Thence of course the Babylonish system was taken to Alexandria; from thence to Rome; and from Rome it has spread like a perpetual cancer more or less into every civilized country of the modern world.

The *standards* adopted under the Babylonian metrology,—seem to have been not a wide variation from the profane Egyptian (the cubit being, according to Dr. Brandis, between 20·866 and 20·670 British inches),—utterly unscientific, so far as regards any known earth appropriate commensurability; entirely unlike, therefore, and totally inferior, to the system of the Great Pyramid; and not known to be preserved in any intentional sizes or proportions for that purpose of the Borsippa or any other Babylonian building. Modern science therefore can say nothing in favour of the Nebuchadnezzarian standards,—while she abhors the employment of their favourite sexagesimal division, in any case where decimals can be used.

Other Competitors.

Hence this very long step in time, that we have taken in order to alight in more advanced periods of human civilization, on a more intellectual building than the Great Pyramid,—has merely landed us in a quagmire of too evidently human idolatrous follies, scientific absurdities, and general mental childishness.

Nor do we improve the opposition by trying anything of earlier time,—for on one side, the Proto-Chaldæan temples were apparently no better than that of Borsippa, may have been much worse, and were certainly less elaborated; and on the other side, no Egyptian Pyramid can for one moment stand comparison with the Great one.

Many of these monuments in Egypt, when duly restored to their pristine state from their present ruins, would be found to have the numerals of the two fives, and the fours belonging to their square-based shapes; but they have not the π figure, nor the numbers 3 and 7, nor the proper latitude, entrance-passage angles, size, material, or internal ascending passages and chambers, with their contents, markings, and structure, nor the proper size and shape of coffer. Their deficiency in magnitude is grievously against them; for while it prevents their vertical height representing by 10^9 the solar distance,—it also prevents the base-sides representing a year in terms of the Pyramid cubit.

And even if we were to make a shorter cubit, by subdividing the earth's semi-axis, not into 10, but into 11, 12, or any other number, of millions, —the numerical terms would cease to be pyramidal, and might not even then be precisely suitable to the measured proportions of any one of the known Pyramids.

Our knowledge of the Great Pyramid may be, and indeed still is, considerably defective,—but no other building comes within its limits of error,[1] which invariably include the natural quantity tendered by the theory. No other building can therefore compete with the Great Pyramid for possessing any of its higher qualifications. It stands the unique building in Egypt and all the early world for intellectual symbolizations, noble thoughts, and physical knowledge pervading its whole substance, to an extent only appreciable to the science of the present and last very few years.

[1] As,—the limits within which its true base-side lengths are supposed to lie, are 9100 and 9170 British inches; while all other Egyptian Pyramid base-sides are stated by Colonel Howard Vyse to be at or under, and generally very much under, 8633 such inches.

The Great Pyramid solemnly alone.

We had previously found the Great Pyramid to be equally unique, or even approaching the miraculous, in the sudden manner in which it appears in history,—the first of all stone buildings ever erected by man;[1]—and in many points, both of excellent workmanship requiring almost mathematical truth, and of absolute height,—still unequalled by anything that has since been erected anywhere; though too 4000 years have passed away—though population has multiplied,—the size of kingdoms grown,—the wealth of nations increased enormously; and though knowledge has long been running to and fro over the surface of the earth.[2]

In short, if the Great Pyramid building had a claim to be considered of something more than merely human origination, regarded architecturally,—then not only 10 times, but probably

[1] See p. 177, chapter 21.

[2] Earthworks of greater horizontal extent than the base of the Great Pyramid, are common enough in America, as well as Europe and Asia. Brick and stone works to the same, or even a greater, lateral extent, may also exist, and are easy to execute. But *height* tests the goodness of a work,—and nowhere else through the world is there any known and published building 5819 inches high.

10^7 times more does it demand to be so considered, when judged of on scientific grounds.

In some things, modern science may have passed beyond the Great Pyramid's power of symbolization; as in computing to any number of decimal places the value of the π fraction; though even then we should remember that the overtaking has only been very recent as compared to the whole interval of 4000 years, and the building, with its necessary mechanical limits, need not be considered expressive of *all* that was in the author's mind. But in other matters, the Great Pyramid is yet ahead; for, granting for a moment the truth of the Pyramid formula,—there is far less uncertainty about the Pyramid sun-distance, as influenced by the probable error of our knowledge of the height of the structure,—than there is about the Astronomical sun-distance, as depending on the different amounts of the several latest determinations of the solar parallax, communicated to the Royal Astronomical Society of London, within the past year.

Still more is the Great Pyramid determination of the Earth's mean density, in advance of all that modern science can do at present. For while my own observations of the coffer show, that a difference of ±·01 in the expression 5·70 cannot be tolerated, and would fall outside all my various coffer determinations of cubical contents,[1]—the last two *scientific* determinations of the same natural quantity differ from each other by 125 times ·01.

These circumstances will probably, even in the opinion of the highest men of science in the present day, justify us in going on to some yet higher considerations.

[1] See pp. 300, 301, chapter 29.

CHAPTER 34.

INTERVENING OBSTRUCTIONS.

UNLESS the author has altogether failed in his duty,—and that little more than setting down facts in order, facts too of a material kind, and amenable to ordinary scientific observation,—the reader will probably now begin to think, and with much ground for confidence, hope, and content, that in the Great Pyramid of Jeezeh we have met with the most remarkable mass of archæological evidence connected with intellectual man—that has ever yet been brought to light in modern times.

In proportion too as discussion proceeds, so does the noble character of our finding continually advance. It is a pearl of great price both in the cause of truth, and for the development of primeval history. A pearl therefore of inestimable value, promising to be worth all the little pearls that have ever been strung on a philosophical thread: and fairly won, after the pains and labour taken, and the rigid as well as open method of investigation adopted.

Yet now, when we are just about to mount upwards to the surface of the ocean of doubts and fears with our prize,—and bear it aloft to the full light of an unclouded day—there hoping to behold some larger use made by others of those qualities which we have taken so long in endeavouring to prove the existence and establish the nature of,—just now rush forward once again determined objectors to both Pyramid theory and Pyramid facts, and do all in their power to impede successful progress. Therefore we, having already adopted the plan of listening to all objections,—have nothing left us to do, than to attend with heed to whatever these latest in the field, these Neoptolemi of the Pyramid question, may have to say. We listen therefore patiently, and trust that not one unanswered complainant will be left behind.

I.

What then says the first?

A scholar is he, nor scrupling to conceal his

scorn of the Great Pyramid as smattering of a mere mechanic; and belonging, for that matter, to a whole crowd of scientific things, all of which he considers low, degrading to a free mind, and with no glorifying impress of classic antiquity upon any of them. He therefore despises them all, and rejoices in logic, metaphysics, and moral philosophy; which triad of learning he proclaims to be the only ethereal food for the soul of man; the only pursuit worthy of the name of PHILOSOPHY; the necessary spinal cord of every school that has ever existed on any part of the earth, even from where the blue waves of the Ægean Sea lap the yellow rocks, to the flashing waters of the Firth of Forth; and a thousand swords ought to be ready to leap from their scabbards, in Edinburgh alone, if only one disparaging look be cast towards her. For 4000 years, as the statement was given forth, has philosophy gone on her ways rejoicing mankind in that magnificent manner. And when a recent University gathering, nothing loath, heard these spirit-stirring words from the lips of their eloquent teacher, heard of school philosophy glorious during the last 4000 years, —there were those in the crowd who bethought them;—'Then where's either the wonder or 'antiquity of the Great Pyramid? Nay, what 'more can your Pyramid measurers say now in 'praise, or even in defence, of those rude, mate-'rial stones?'

Not much more indeed, in the way of absolute age, than what we have said already; but the real question is, can so much be truly stated for philosophy?

So therewith we wrote to the learned lecturer in the zenith of his popular fame, requesting to be informed what were the text-books in moral philosophy, logic, and metaphysics 4000 years ago, and who composed them. Whereupon that at heart honest and true gentleman instantly sent back a most courteous answer to confess 'that he 'had made a slip; he had only wanted to express 'a vaguish number, but had taken one, it would 'seem, that was too high; and he had no inten-'tion of maintaining anything upon it.'

II.

Advances then a second opponent cautiously; beginning with 'whereas you have been discover-'ing in the Great Pyramid, with charming suc-

'cess, so many things known to science very well 'of old,—we are hoping now, that you will before 'long discover something *new* there.'

This gentleman we proceed to assure immediately, that we have never looked on the Great Pyramid, as standing, or ever intended to stand, in the place of modern science. Modern science for its own purposes is unequalled; and except for its extreme development in present times, the Great Pyramid would have remained still uninterpreted and unappreciated. Indeed, if there is much more still to be discovered in that most ancient monument, appearances and the course of all inductions hitherto, seem to say, that we can only expect to ascertain it securely and certainly from the still further progress of modern science. No one therefore wishes so cordially the rapid and independent development of all true science, as those who have laboured in recent years on the interpretation of the Great Pyramid.

'Then if you neither *have* discovered, nor ever '*will* discover, anything scientifically new in the 'Great Pyramid,—what is the advantage, or ob'ject, of looking so much to the Great Pyramid? retorts the questioner with severity!

To whom however we reply—

Even if no actually new facts in terrestrial physics should have been discovered, or be discoverable,—yet, *first*, the giving of even one improved determination of an older phenomenon, is invaluable for suggesting thought. *Second*, the offering of a collective system of metrology, eminently suitable to man, and more scientific than anything he has ever had before in actual and material use, is more momentous still. And *finally*, the extensive proving that many facts only recently discovered by science, had been intentionally inserted, symbolized, or memorialized in the Great Pyramid 4000 years ago,—appertains to a new fact of a much higher order, or one connected with the actual history of the human race in primeval ages : and which history, in certain of its stages, none of the discoveries of modern science either touch upon, or have it in their nature to be able to illuminate, even in the slightest degree.

III.

Now history is always allowed to be important, but to get at the soul of history, and in primeval times, near the epoch of the origination of intellectual doings on the earth,—that surely is of surpassing importance to all men at present living.

How often do our writers, yearning for more of the spirit and less of the letter, quote from one of Britain's most eloquent sons,—

> 'How cold is all (book) history, how lifeless all (literary) 'imagery, compared to that which the living nation writes, 'and the uncorrupted marble bears! How many pages of 'doubtful record might we not often spare, for a few stones 'left one upon another.'[1]

They sighed, those worthy souls, for only a few stones left one upon another, as a better principle of investigation, than doubtful essays on the long-departed past. And now when the Great Pyramid offers them thousands of stones standing regularly, systematically, scientifically, and with a mighty purpose of profound and deep-ordered meaning, one upon another,—forming a mass suitable to being compassed in the neatest geometrical definition, and capable also of being extended by its own not altogether latent formulæ into the mensuration of the earth and the

[1] Ruskin's *Seven Lamps of Architecture.*

meting out of the heavens,—now these enthusiasts suddenly forget their late aspirations, their ardent desires, and hide themselves far off from the neighbourhood of the almost speaking stones of the antique Pyramid.

IV.

Alas! they have been frightened. A medical Baronet has come down to the Royal Society of Edinburgh, and has violently denounced Professor Smyth's published measures and calculations of the Great Pyramid together. The said Baronet 'of brilliant and diversified gifts,' in the *Medical Times*, had already, according to that paper,[1]

[1] From the *Medical Times*, of London; March 28, 1868:—

'SIR J. Y. SIMPSON ON THE PYRAMIDS.

'It is characteristic of the brilliant and diversified gifts of the 'great Northern professor, that he should have taken in hand the 'demolition of a curious notion about the Great Pyramid of Egypt, 'which has been promulgated by Professor Piazzi Smyth, and has 'found favour with many religious persons. That notion is, that 'the Pyramid was constructed under Divine guidance, to show to all 'time a correct standard of weight and measure, and that the coffer 'contained in the central chamber is an inspired measure of capa- 'city, and the base of the Pyramid an inspired measure of length, 'having a definite relation to the earth's polar axis. Sir James 'Simpson shows, in a paper read before the Royal Society of Edin- 'burgh, that the Pyramid has all the characters of the huge sepul- 'chral monuments scattered over the earth; that the coffer in the 'King's Chamber was a sarcophagus; that it contained a body till 'despoiled by the Caliph El Mamoon about 1000 years ago; that completely demolished the scientific metrical theory of the Great Pyramid, even in a way that was 'quite a treat' to see,—in a paper in the 75th Number of the *Proceedings of the Royal Society of Edinburgh*;—but on the 20th of April, at another meeting of the same Society, that great medical man transcended all his former efforts, and beginning upon a copy, in form of a table, of the coffer measures contained in our pages 297 to 301, —declared, as we are informed, that *all* its figures were wrong, and wrong in ways and modes which made the means of the observations worse still. Professor Smyth also was declared to be hopelessly given to the 'erroneous method of taking '*means* of different measurements of the same 'thing; such a mode of procedure was denounced 'as childish; it was a species of mathematical 'aberration, and it ran through the whole of 'Professor Smyth's book. . . . Sir J. Y. Simp-

'it is irregular in form; incapable of being exactly measured, and 'hence no standard of measure. In fact, any one who desires a 'treat in seeing how a thorough "craze" is melted away before 'common sense, should read Professor Simpson's paper. The kind 'of argument of Professor Piazzi Smyth to show that the Pyramid 'was built by Noah, is one which we thought peculiar to the medical 'profession:—Noah was a preacher of righteousness; a just system 'of weights and measures is righteous: *ergo*, Noah built the Pyramid!'

'son had not heard a single sound man say 'otherwise than that the whole of Professor 'Smyth's theory about the Pyramid was a series 'of the strangest hallucinations, which only a 'few weak women believed, and perhaps a few 'womanly men, but no more. He had talked 'about it to a great many engineers, mathemati-'cians, and others, and he found them scoffing at 'and despising it.'[1]

The fact that Professor Smyth was left alone before the Society, to bear the above torrent of invective—will, in a future day, be looked at in the same manner in which we now regard the records of those old courts of 'justice,' where learned judges formally condemned both men and women for allegedly *proved* acts of necromancy, witchcraft, and other really impossible matters; and no one said a word against it.

For how can injustice go further—than that when the method of means, or of taking the *average*, is used in every scientific, statistical, and commercial operation throughout the kingdom and the world,—that the application of it to

[1] From the Edinburgh *Daily Review*, *Scotsman*, and *Courant*, of 21st April 1868.

measures taken on the Great Pyramid,—is alone declared to be erroneous, equivocal, childish, and a mathematical aberration? No problem in practical astronomy is ever carried on without it, and yet it is to be held a sin or a crime in Great-Pyramid explanation! It is surely to be hoped that Sir J. Y. Simpson will publish his ideas in full; for at present he has only given his assertion that the method of means or averages is a mathematical aberration; and the mathematical world will expect, that he who appears in the *Medical Times* as 'the great Northern professor, of brilliant and diversified gifts,' and who has, in that Journal's eyes, demolished the scientific theory of the Great Pyramid, largely too by employing this peculiar engine, will either prove his assertion, or——

But then Professor Smyth's published measures and other calculations besides merely the means, were nearly all declared on the same occasion to be wrong. He has sought to find where, but cannot discover. He has gone through all the coffer multiplications and divisions whose results were exhibited to the Society at the time,—and can find no error in them, to the last place of figures represented. And he laid on the table of the

Society a printed list of errata in the volumes of his *Life and Work*, containing all the *corrigenda* he had been able to discover therein up to that date, and they are reprinted in the Appendix to the present book. There may be some stray errors, as there are in all large books of any private author, and notices of them, when discovered, will be gladly received by him;—but where are those nearly universal errors which were denounced to the Royal Society of Edinburgh, or errors to an extent which make it futile to found any theory upon those measures of the Great Pyramid?

We have not shrunk from giving these denunciations by Sir James Y. Simpson, Baronet, at length,—in order that if they are true, he may gain due honour therefor; receive the thanks of all who are interested in the progress of Pyramid knowledge, and even achieve the fame of St. George for slaying the very dragon of Egypt. But meanwhile we have a strange doubt as to the whole attack. Every numerical expression, we have re-examined as yet, has proved correct. Even so pointed a case, as Professor Smyth being dragged up in italics, and with a (*sic*) in parenthesis, in Sir J. Y. Simpson's 'Proceedings' paper, as having used the vulgarity of '*extraordinary* 'convenient,'—was proved before an open meeting of the same Royal Society of Edinburgh to be the Baronet's own expression,—while his mis-statements and perversions of general meaning were shown to be abundant. In fact, we rather suspect that the whole attack, or series of attacks, carried on through several months past by the above chief of medical men, on both the Metrical theory of the Great Pyramid and Professor Piazzi Smyth, is an attempt simply and shortly to *stamp them out*, in the style so much lauded just now in veterinary circles from its recent practical success in the case of the cattle-disease.

In such an operation, of course, no great nicety of treatment need be looked for by the parties on whom the stamping is to be performed.

Hence perhaps it may be, that besides some reckless assertions of arithmetical and grammatical errors which do not exist, the stamper-out gave his own *ipse dixit* to the Royal Society, Edinburgh, that engineers, mathematicians, arithmeticians, and others, scorned the theory, scoffed at and despised it. For, as a friend has since remarked,—

'What an ill-mannered thing to tell *that* to the Royal 'Society of Edinburgh as a reason *to them;* as if they were 'not necessarily themselves the *élite* of the professionals 'whose opinion he thus asked them to adopt;—as if they 'were not met expressly to judge *for themselves*, from his and 'your discussion then and there. A jury would not feel 'flattered if counsel for the plaintiff urged as an argument 'to move them, that all the people outside the court thought 'defendant guilty.'

But when once the idea of 'stamping out' is indulged in, the stamper is not particular on whom he stamps, so long as he does stamp far and wide over all the obnoxious region; wherefore, not content with denouncing, without proving, Professor Smyth's book as containing 'more extra-'ordinary hallucinations than had appeared in 'any other three volumes published during the 'past or present century,'—Sir James Y. Simpson went on to deride all who believed in the theories and discoveries mentioned in that book; and declared such persons to be—'only a few weak women, 'and perhaps a few womanly men, and no more.'

V.

Of course a scientific discussion can hardly be satisfactorily conducted amongst *men*, after one of the parties has begun to indulge in such phrases as the above. But a young lady may always be heard with advantage and pleasure; and the newspaper containing the so-called Pyramid discussion of the 20th of April, having reached a well-ordered home,—a daughter of the house, who had both seen the Great Pyramid of Jeezeh, visited its darksome interior, and also read most of the recently published books on that and other Egyptian subjects,—took up her pen and wrote off next morning, amongst other very apposite remarks:—

'Sir James Y. Simpson does not seem to have proved a 'single objection to the printed Pyramid theory, with the 'exception perhaps of a few printer's errors; and his whole 'attack shows anything but a scientific spirit; while his scorn 'of women, and his epithets on those who differ in opinion 'from him and his friends, are much to be lamented for what 'they reveal, beneath his smiles, of his inner nature. Be 'it observed, however, that he involuntarily gives women 'credit for more learning and science than he possesses him-'self,—if they are capable of comprehending and appreciating 'for themselves such deep questions as those involved in the 'Metrical theory of the Great Pyramid. For *he* gets opinions 'on it from other persons, and retails them to the Royal 'Society of Edinburgh; but some women who have had time 'and opportunity, read, think, and reason on these matters 'for themselves, and in that case they ought not to be termed '*weak*.'

'It is a pity Sir James Y. Simpson should waste his time

'in trying to disprove a subject which is not at all in his sphere, 'and about which he shows himself so ill-informed, having 'neither had time nor opportunity thoroughly to examine the 'grounds and foundation of the Pyramid and its teachings; 'and indeed, unless he goes to work with a more disinterested 'and truly philosophical spirit, I fear that the recent prescrip- 'tion exhibited to him, of ascending and descending the 'Pyramid itself, and measuring on every side and every 'arris-line of the structure,—will not cure him. Truly of 'him, on the present topic, may it be said, and with an *extra- 'ordinarily* close application—

'"*He is proud, knowing nothing, but doting about questions 'and strifes of words, whereof cometh envy, strife, railings, 'evil surmisings.*"—1 TIM. vi. 4.

'The best advice that can be given to Sir James Y. Simp- 'son is,—"*to study to be quiet, and to do his own business, and 'to work with his own hands.*"—1 THESS. iv. 11.'

CHAPTER 35.

CONSIDERATIONS IN CHRONOLOGY.

WHY, so calm an inquiry as the study of ancient masonry, and the scientific measuring of the shapes or sizes of quiescent stones—still resting where they were deposited by their original workers, but in days removed from our present worldly scenes of human strife and trouble by 4000 long and eventful years,—should have been greeted with the apparent hatred, and attacked with the persevering vindictiveness, shown of late towards investigations into the Great Pyramid,—seems on many accounts passing strange. And yet, on reviewing the quarters whence the opposition has chiefly come, the matter nearly explains itself.

Rationalism is the prevailing tone of the day, whether in science, or religion, or both together. Even with regard to the Book of Inspiration, the question on too many sides is no longer, 'What 'readest thou?' but 'What thinkest thou? Are 'these the sort of words, sentiments and ideas 'which *should* be found in a book written under 'Divine inspiration? Nay, if the book do not fully 'meet thy *a priori* ideas of inspired wisdom, cast 'it from thee!' While as regards many walks of science, the dogma recently stated there is, 'We may 'be sure that God does not teach by miracle, any-'thing that the unaided intellect of man can find 'out;' and then there are certain *savants* who cap that with—'Give us only liberty to use plenty of '*time* in our theories, and we will show you how 'the mere brain, or natural organization of man, 'could have worked out for himself, everything 'which he at present possesses.'

So extensively does this idea of the mighty things which man can accomplish by merely having plenty of time to work in, prevail, that before most of the scientific societies in the land, we believe no other hypothesis is allowed to appear. And each successive difficulty connected with human history, and the present condition of any of the sciences, is got over,—by merely pushing the real origin of man still farther back by some thousands, or hundreds of thousands, of years, than it had already been pushed back by similar theorizers, beyond the Bible-reputed date of his creation.

Of course we are told by those who are naturally inclined to such a mode, that there is no harm in it; it is only what they are entitled to do; for time, they hold, is something you may be prodigal of and waste to any extent you like. '*I* 'know of no moral or religious truth,' says a noble Duke, writing recently on the antiquity of man, 'which depends on a *short* estimate of man's anti- 'quity. On the contrary, a *high* estimate of that 'antiquity is of great value in its bearing upon 'another question, much more important than any 'question of *time* can ever be.'

Again, his Grace testifies for the science of the day, that 'there is a weight and concurrence of 'authority in favour of a *long* chronology;' and for himself, 'that for his own part, he sees 'no reason to be jealous of the conclusions of 'science in this matter. The question is after all 'a small one. It is a question of a few thousand 'years more or less; and thousands of years are 'as less than seconds in the creative days.' And again: 'The difficulty of reconciling the narra- 'tive of Genesis with an indefinitely older date, 'is a very small difficulty indeed, as compared 'with the difficulty of reconciling it with a very 'limited destruction of the human race.'

Time is thus indicated as a something which the titled author has a marvellous power, and despotic control, over. He disconnects too, after an arrangement of his own, the question of time, as regards man's antiquity, from either his origin, or primitive condition; and then, in that compendious form, carries back the time of man's creation or introduction into the world, to certain eras of geological history and preparation of the globe, when there was no *time-absolute* (as within the historical period, and which, according to him, only goes back to 2000 B.C.)—but only *time-relative;* where the sole limit appears to be, that you are not allowed to assume actual infinity, though no one can chide you for taking as many myriads of years, as may suit your convenience or purposes.

Check on Prodigality of Time.

Now, to our humble apprehension, Time is one of the God-created four,—Matter, Force, Space, and Time; whereof Time is just as valuable as any of its compeers, and as vastly above the power of man to supplant, appropriate, or do away with; and all of them must have been created in perfect economy as well as power, and completely foreseen in all their characters, extent, and bearing, by the infinite wisdom of their Creator. Wherefore we ask, why all this contempt of Time, alone?

'Because,' says a well-meaning Scripture believer—'give 'us only plenty of time, and then we can derive all the exist- 'ing varieties of men, as Negro, Caucasian, Mongol, and Red 'man from a single original pair: and that is agreeable with 'the Biblical statement of how man was created. It is there- 'fore for the actual support of the Bible, in its more important 'relations, that we demand such long periods of time since 'the creation of the first human pair.'

We rather fear, however, that trying to support the Bible in that manner, is too near an approach to the spirit of Uzzah, when 'he put forth his 'hand to the Ark of God, and took hold of it; 'and the anger of the Lord was kindled against 'Uzzah; and God smote him there for his error.'

The method, moreover, of demanding the creation date of man to be pushed back enormously, has been found only too congenial for those who openly profess disbelief in the inspired character, or Divine truth, of the Scriptures; who boast that there is no trace throughout all nature of any of these interferences with natural laws which the Bible alludes to; and who,—caring little about origin from one pair, or shoals of pairs, and one place, or many either simultaneous or successive places, of creation,—direct all their attention to the *kind* of origin they find for man; and rejoice to place that origin as low in the scale of animated being as possible; so that some would trace all present men from an ancestor lower than that of the lowest savage anywhere existing; some would bring him from the far lower depths of the nearest anthropoid animal: and some from the vastly lower depths still of the remotest monad, or atom, of any kind of physical existence whatever.

This seems rather an untoward mode for deriving the Godlike mind of man; but, with only plenty of time, these 'advanced-thinkers' declare that they can do it. That is, they can show, on paper, how man could have risen from any of these states, to his present condition,—merely by his own animal organization exerting itself always in one direction, for a sufficiently long period of time; but in accordance with natural laws, and in modes completely amenable to science, and wholly without the range of interference of a personal Creator.

So successful too has this paper demonstration been, to many minds,—that its cup of glory has somewhat overflowed of late; and many well-meaning men who began with demands for only a little more time, and some also who were even scandalized at any more time, than that allowed by the Bible dates, being asked for,—are found now freely consenting to this other Inspiration subversive proposition,—viz., that man's earliest position on earth must *necessarily* have been infantine, rude, and savage to the lowest degree.

If we ask them, why that low beginning *must* have been :—they answer—

'Because there was then as little as possible to be done by 'the Creator at, or after, the creation of man, in a manner 'which no modern scientific society can, or does, allow; but 'as much as possible left for man to work out by his own 'powers on the principles both of rational science, our pre-'sent modes of improvement, and progressive development 'methods, whose truth and sufficiency all *advanced-thinkers* 'admit.'

Needless to point out what a limiting of the power of the Creator is here implied; what a dropping out of view some of the most crucial statements in the Bible touching the Inspiration divinely afforded to the first Man; what a case of improperly putting forth the arm of Uzzah to assist Him whose power is infinite. Needless we say for us to point out these things here,—because it is neither our responsible duty so to do; nor are we contemplating any discussion of the case on theological principles, or even Biblical grounds. All we are professing to do, is merely to compare the general resultant of all our Architectural and Archæological securely ascertained testimony, with,—on one side, the doctrines of rationalism; and on the other, the leading and simple statements in Scripture history; taking this latter for the time, merely as an hypothesis to be tried; or precisely in that form in which, however innocent, and though perhaps just and true in fact, no first-class scientific society of the country will allow any appearance to be put in for it.

The Science-interdicted Hypothesis tried.

First then, let us remind, that our preceding chapters 22 to 32 tend to show, that although there may have been generally advancing school-science in the world during the last 2000 or 2500 years;

—and that such species of knowledge must therefore have been at a far lower position at the date of 500 B.C., than now amongst ourselves, or at a positively very low level indeed,—yet, 1600 years previously to that again, or in 2170 B.C., there were built into the Great Pyramid symbolizations of astronomical and physical data, expressed to such surpassing accuracy, as to be rather in advance of, than behind, the science of the present day.

This great fact, or general induction from many facts, instantly strikes at the root of one rationalistic doctrine; viz., 'that man never forgets any'thing good or important that he has once ascer'tained; and that his course is therefore necessarily 'always an advancing, gaining, and triumphant one.' That doctrine we say is overturned by our broad facts, because man evidently did fall away, somehow or other, from the knowledge which he had at the time of building the Great Pyramid,—and which knowledge he has not recovered, after ages of slow and painful climbing upwards from an intermediate state of ignorance, until within these last few years.

'Well!' some honest *advanced-thinkers* may confess, 'granted that human progress *has* met with occasional re-

'verses; these are mere accidents, occurring from war, pes-
'tilence, or other calamities. But the really important ques-
'tion is, how was that old knowledge at the Great Pyramid
'date acquired; was it slowly on our rationalistic methods, or
'suddenly by Divine Inspiration?'

Precisely there, we too allow, is the main question to be settled.

On rationalistic principles, the knowledge of astronomical and other sciences in the Great Pyramid, could hardly have been acquired by man for himself under a previous career of something like 4000 years, even if he had then started,—at the beginning of that very early time, or 8000 years ago,—from a state of general knowledge and civilization equal to what was enjoyed in Eastern countries in the time of Joseph. Nor could it probably have been attained under 40,000 preliminary years, if man had had to start, and plod on upon his course unassisted by a higher power, from a state, as some maintain it, rather lower than that of an Australian savage. Nor in 400,000 years, if ever, supposing that he had had to work up from a gorilla or monkey condition; and how long from a monad, we will leave the *advanced-thinkers* to say for themselves.

But we need not now attempt to discriminate

between, or choose any one of these origins, rather than another; because the last 4000 years of each of them, if successful in producing the Great Pyramid in the end, would be nearly or precisely similar. For these last 4000 pre-Pyramid years, therefore, on any rationalistic system, there should have been, and there should now be, the remains of innumerable stone buildings,—gradually approaching the maturity of excellence seen in the Great Pyramid;—such monuments should indeed be in exceeding profusion, scattered up and down in various countries of the earth; and containing within them interpretable proofs of constantly advancing science, knowledge, and power.

This is what *should* be, and oh! how abundantly, according to rationalism; but scientific examination of the actual facts existing in the world, whether the reader may take such examination from Fergusson's *History of Architecture*, or from our own previous pages,—declares, that excepting some problematical and not very important structures [1] of a few centuries earlier only, —the world has no material and contemporary record of intellectual man, earlier than the Great

[1] See page 170, chapter 20.

Pyramid; *i.e.*, no earlier architecture and no earlier science traces, even of any kind. The Great Pyramid opens human architecture, not with a gradual beginning, slowly growing up from a series of almost invisibly small and weak attempts, improving through countless ages, but with a sudden burst of height, breadth, majesty, science, and excellence, to an extent which is for all practical purposes even perfection itself.

This extraordinary difference from the *rationally* expected result, is not merely a defeat; it is a catastrophe approaching to annihilation for rationalism. Not indeed of our doing, or even of Mr. Fergusson's either; but simply by the actual facts of the earliest history we can get at,—and as taken account of by modern scientific examination of the only undeniable evidence, viz., *contemporary* records,—being antagonistic to the doctrine.

We have shown, at pp. 183-332, what the Great Pyramid, with its science, was at the time of its building, or in 2170 B.C.; and advanced-thinkers must now, either prove why there is a perfect vacuity of monuments, in place of their required *plenum*, before the Great Pyramid era; or they must cease to advocate rationalistic philosophy, as able to explain the actual course of man's real history under the sun,—whatever they may make it do for a mere paper system of a Utopian creation.

Language.

Some very strenuous efforts have been made recently, to cry up the newest forms of the science of philology, as composed of glottology and mythology;[1] the leading principle of which, is nothing but an extremely *advanced-thinking* endeavour to show how man could make language, out of mere power of animal sounds, in sufficiently long ages,—in place of having had it at once given to him complete by the Almighty. That such a hypothesis should be tried on paper, we presume will not be objected to; but that, because it has apparently succeeded there, with the gift conceded of unlimited time,—which man has not in his power really to make free with, —that therefore certain scholars should boast they have proved that their hypothesis was both

[1] 'Glottology, the science of elementary vocalizations, their origin, 'significance, and combination historically into speech: and Mythology, the science of the primitive, transmitted, and perhaps 'organic imaginations and conceptions of the human race.'

the law of nature and the fact of man's early history—is what mere practical observers must positively demur to.

Wherefore, although those advanced scholars write triumphantly amongst themselves, that they are already, by means of their favourite,—and which, if the youngest, they say is the most scientific, of all the sciences,—already twining their hands into the mane of back-rushing time, and compelling the monster to stand, and extracting from her some of her obscurest secrets :—also

'that they find their science promises to be a calculus of as great potency for solving problems of the human past, as geology has been for solving those of the *pre*-human past, but of still greater exquisiteness and complexity :—also

'that the evidence of philology in language is irrefragable, and is the only evidence worth listening to with regard to the pre-historical periods :—

'that philology affords the most conclusive evidence of all the sciences yet employed in the inquiry, for showing that the articulate voice of man has been sounding in the world during vast, though indefinite periods of time :—and

'that those who have studied the growth of

languages, and the mysterious laws by which that growth is regulated, are lost in conjecture as to the lapse of time which may probably have been required to account for the wonderful creations of human speech :'—

though all this, and very much more too is asserted in vigour of grammatical phrase, and deeply resounding periods of apparently transcendent philosophic wisdom, inimitable except by equally great scholars with those who have pronounced them,—still we do not see that those gentlemen have proved a single philological fact in primeval history, by earlier, or even so early, contemporary evidence as the Great Pyramid. Nor do we apprehend that all their learned schemes, are representing anything more than merely modern lucubrations of ingenious building up, on paper ;—and no higher position can well be allowed them, until they have discovered a method of raising their theoretical predecessors out of that woful pit of perdition whereinto they have fallen, by having no real monuments to show for all the ages preceding the rise of that august witness to modern times, of what really did take place 4000 years ago ; viz., the Great Pyramid.

The Deluge generally.

'But my theory failing does not oblige me to 'believe in yours,'—exclaimed a sufficiently clever advanced-thinker to us recently. 'You say,' he went on, 'that nothing but the admission of 'Divine inspiration to man, and of interference, 'so far, with the order of nature, will suffice to 'explain the observed Great-Pyramid facts. But 'these facts do not suit the Bible history either; 'because according to that,—and if Adam was 'really inspired, and his descendants endued at 'once with all the wisdom and capacity thereby 'implied,—then there were some 2000 years of 'intellectual life before the date of the Great 'Pyramid. Where then are all the monuments 'of that long period?'

Now this is indeed a question to which we would desire to draw the earnest attention of all. The Bible does no doubt place the creation of man, and the breathing into him, by Almighty power, both the breath of life and apparently much spirit of wisdom, language, religion, and many of the arts of industrial life,[1] say some six

[1] Amongst these may be reckoned the pastoral pursuits of Abel; for although they are considered by some persons as the occupa-

or seven thousand years ago; for we are not competent to interpret the numbers perfectly there; but at all events nearly two or three thousand years before the building of the Great Pyramid; and we believe the account true, in all its largest and most understandable features. But then the Bible also recounts the terrific story of the Deluge, as having occurred only a few centuries before that Great Pyramid-building date; and having also, both removed all traces of the previous population, and reduced mankind again,—so far as mere numbers were concerned,—to the condition of the first created family, with all the earth before them to people with their descendants to be.

Take therefore the Biblical account of the Deluge, as a necessary and true part of the Biblical history of man from his creation, downwards to ordinary historical times,—and all the main

tion of a primitive savage,—they could not at once be put into execution at any one epoch of time, unless animals had been *prepared beforehand* in their nature and inward parts willing and suitable to become the domesticated animals of man. In the earlier of the *Egyptian* tomb-paintings, though that Cain-following people were abundantly provided with the fattest of oxen, sheep, and goats, yet they were always striving to domesticate large antelopes, apparently in place of oxen, and hyænas in place of dogs. But all those long-continued human efforts totally failed in adding either one or other of those wild beasts permanently to the stock of the animal companions and servants of man.

features of both the Great Pyramid architecture and Great Pyramid symbology—all the main story in fact which the Great Pyramid has to tell, is more than ever in favour of the Scriptural version. And the Great Pyramid is a monument not of the creation and first inspiration of man, but of the Deluge, and some of the first steps connected with the spreading of population some centuries afterwards; possibly, or even probably, after the dispersion at Babel.

There are no doubt serious difficulties in the way of understanding the precise interpretation of every Bible detail connected with the Deluge; though that event may be more simple, as it is also more recent, than the creation of man; but, as to the general fact of a miracle of such an inundation order, with the *purpose* too of removing an apostate population from the surface of the earth, and with an unprecedented destruction of the human race in effect, and as having taken place somewhere between 2300 and 3300 B.C.—the admitted texts of both Old and New Testaments are most positive and unequivocal. Prophet and Apostle, and our Saviour himself, all testify to the general leading fact of a sudden, and anti-natural watery destruction of all men but one selected family; and there is no possible cause, compatibly with honest belief in the Bible, for either pushing back the date of the Deluge to an indefinite number of myriads of years ago; or making it partial amongst a small portion only of mankind; or again for giving to the visitation anything else than a miraculous and nature-interference characteristic.

Yet some men do any or all of these three things, and actually profess (in a manner we cannot understand, though it is becoming a prevailing feature of the present day) to be firm believers in the Bible all the time. They claim to act in all things as the best of Christians, to hold the Bible as their Sacred book,—and yet delight to turn into ridicule any one who believes in the personality of Noah, or the reality of the Deluge at anything near the Biblical recorded time.

Herein indeed, as connected with attacks on that Biblical Deluge and its date,—is another of the strongholds of advanced-thinkers; though it is one which may now be visited and tested, as it never has been before, from the vantage-ground of the Great Pyramid. Rationalism, in all her examinations of nature and the world, can discover no traces of the Biblical Deluge, and therefore disbelieves its having ever occurred. Her followers, too, had rather an easy victory put into their hands on one point of the discussion, by the ill-judged ardour of certain Scripture-following scholars of the last generation or two. For these, being far too prone to think the Deluge must have been material and natural, rather than miraculous,—invariably hastened to seize on every case of a geological elevation of the bed of an old sea, or any traces of accumulated water-wearing action, as proofs positive and direct of the Noachian Deluge.

Rationalism therefore triumphed gloriously then, when geology showed the true nature of these particular phenomena, and the manner of their occurrence without the aid of any universal flood. But now that the same party has gone on since then to declare, and even boast itself in declaring, the further supposed geological discovery, —or that, so far as the surface of the earth is concerned and all the mutations it has been exposed to through countless ages,—modern science can find 'no sign of a beginning, no prospect of an end,'—why, the advanced-thinkers have fallen into the worse scientific error of the two.

A killing error; because, not only does the calculating natural philosophy of the present day, declare that such a uniform system in nature is just as impossible as perpetual motion in mechanical art; but now, the metrically scientific examinations which have been recently made of the Great Pyramid, combined with traditions of ancient, and calculations of modern, astronomy—tend to throw a peculiar light upon the very epoch of the Deluge; and prove that *time* is no more to be trifled with in its grandeur, when connected with the Biblical history of intellectual man, than any other of the particulars with regard to him contained in the Sacred book.

Date of the Deluge.

We have already set forth the confirmation which the mere fact of the existence of the Great Pyramid, in its size, excellence, and unpreceded appearance gives to mankind having then,—*i.e.*, in, or near to, 2170 B.C.,—only recently emerged from a period of very small numbers, but with supernatural assistance; and now we have to detail the further and more particular testimony as to the date of the terrible catastrophe itself.[1]

Recent exploration of the Great Pyramid, has shown that the distinguishing feature of its astronomy, was the simultaneous observation of an equatorial star crossing the meridian above the pole, when a certain chosen circumpolar star, viz., *a* Draconis, was crossing the meridian below the pole.

It was this system which gave to the Pyramid the Pleiades, with all their traditional associations of grateful memories, from 2170 B.C., and for a long time downwards after that date. It was that system also which, according to calculations based on modern astronomy, would have given to any man observing stars on that Great Pyramid principle, in antediluvian periods, or from 3400 B.C. and upwards,—the stars most maligned by tradition as connected with times of evil fate, viz. the stars of Scorpio.[2] And it is finally the same system, which must have given to the middle of the intervening time, say 2800 B.C., as the domi-

[1] See the opening pages of W. Osburn's *The Religions of the World*, for some apposite details of the earliest form of religion practised amongst the ancient Egyptians.

[2] See *Life and Work at the Great Pyramid*, by C. Piazzi Smyth, vol. iii. p. 492.

nant constellation of that age, *Aquarius*; or a constellation always connected in the minds of men with ideas of a flood of some kind or another.

Not that there is any sort of virtue, good or bad, in any of these stars, enabling them to produce physical effects upon the earth, in and by their own power; but that certain historical ideas were attached to them either by names or descriptions of figures in appropriate actions, as memorials to men, long before the days of classic science began; so long ago indeed as to have extended into the traditions of almost all lands, and to have been already attributed to certain branches of inspired teaching in the early days of the world, uttered for prophetical purposes closely connected with revealed religion.[1] And now the Great Pyramid gives us the key to some, at least, of these most primeval asterisms; and makes the date of the Deluge, by the manner in which it singles out Aquarius from all the other zodiacal constellations, to be as surely very near to 2800 B.C., as the date of the Great Pyramid's own building is close to 2170 B.C.

In point of time then, behold the Bible date

[1] See *Mazzaroth, or the Constellations*, by the late Miss Rolleston.

singularly fulfilled. For 2800 B.C. is close to the mean of the dates of the two chief versions of the Scriptures, the Septuagint and Hebrew; whereon every Church throughout Christendom sees its sons utterly divided in opinion, as to which of the two copies is the more trustworthy. Hitherto divines have generally held either to the 3300 of the Septuagint, or the 2300 of the Hebrew, and have accused the opposite version of having been either subtracted from, or added to, by the whole amount of the difference. But now the Great Pyramid offers its well-authenticated and remarkably preserved original testimony, declaring the truth to be nearly in the middle between the two; and gently reminding, that if man had trusted the catholic and whole spirit of all the Scripture versions which have come down to our times,—he would never have been sensibly far from the truth.

Hence the final result of all our research into the antiquity of Man, by the most certain, nay only certain, plan of contemporary data, is,—

First, an admirable testimony to the truth of the Scriptures, in the dates therein given, from historical times up to, and including, the date of the

Deluge; further than which the Great Pyramid does not enable us to go: and

Second, that no traces of intellectual man, earlier than 2800 B.C., will be found anywhere.[1] Any traces earlier which may ever have existed, would seem to have been purposely made away with by a *supernatural* catastrophe, into whose *modus operandi* it is in vain for man to inquire by aid

[1] We may be excused for pointing out here, how the date of 2800 B.C., relieves some believers in the supposed Hebrew Deluge date of 2300 B.C., from a difficulty which they had met with, and honestly confessed. In George F. Chambers's *Astronomy*, Oxford Clarendon Press Edition, p. 42, we read—

'The earliest record we possess of an occurrence of this kind (a 'close apparent conjunction of several planets) is of Chinese origin. 'It is stated that a conjunction of Mars, Jupiter, Saturn, and 'Mercury, in the constellation *Shi*, was assumed as an epoch by 'the Emperor Chuen-hio, and it has been found by MM. Desvignoles 'and Kirch, that such a conjunction actually did take place on 'February 28, 2446 B.C., between 10° and 18° of Pisces. Another 'calculator, De Mailla, fixes on February 9, 2441 B.C., as the date 'of the conjunction in question; and he states that the four planets 'named above, and the moon besides, were comprised within an 'arc of 12°, extending from 15° to 27° of Pisces. It deserves men-'tion that both the foregoing dates precede the Noachian Deluge. 'It can therefore only be that the planetary conjunction in question 'was after-ascertained.'

The last two sentences are evidently now rendered unnecessary, by the mean Biblical date and the Pyramid date both giving the Deluge earlier by several centuries than 2446 B.C. But it still remains to be proved, whether the Chinese statement may be regarded as an old contemporary record, or a modern composition. We fear the latter: though always expecting that some important chronological discoveries may still be made amongst that very ancient and most incomprehensible people.

of any or all the varieties of modern science, which is essentially *sub-natural.* While of all immediately subsequent traces of intellectual man's handiwork, only that one has well survived to the present time,—where the Divine favour was extended to the builders thereof, both in method and design; for a purpose never yet distinctly revealed in words, though it is now apparently, like the fulfilment of an ancient prophecy, gradually coming into view.

Precisely too as more and more of the intended end is accomplished, so will man be allowed and enabled to see more clearly what were the original aims and intentions of the whole design.

Sadly may the Great Pyramid be now already broken and injured beyond repair; and social opposition in our own community may run strongly, nay violently, against any worthy national effort being put forth, even for the monument's full mensuration, before decay proceeds much further. But we may nevertheless rest perfectly assured, that enough of it will still be left, to show,—and when the right time comes, to prove,—that everything was foreseen, and pre-arranged in every particular, from the beginning.

Of other antiquities, and those too which up to the present time have chiefly occupied every Antiquarian Society,—an eloquent author has written, in apparent despair at such remains disclosing only traces of confusion with lamentable want of method, experience, and skill,—

> 'We must give it up, that speechless past; whether in fact 'or chronology, doctrine or mythology; whether in Europe, 'Asia, Africa, or America; at Thebes or Palenque, on Lycian 'shore or Salisbury Plain: lost is lost; gone is gone for 'ever.'

It may be so, with all old works of merely human origination; but it will not be so with the still older Great Pyramid, *if* a Higher Power has otherwise willed it. And certainly from the moment of man having addressed himself to examine that particular monument with respect, earnestness, and science, it has begun to yield a series of results, which completely separate it from all that is so comprehensively alluded to in the thrilling confession of human insufficiency, just quoted.

APPENDICES.

'Again, if it were said that *that* (only) is a fact which our senses 'perceive, the question must be asked, *Whose* senses? One man 'watches the stars all the night, and sees them describe (apparent) 'circles about the pole; another looks at them carelessly and at 'intervals, and sees no circles.—And thus what is matter of theory '(or even total disbelief), to one observer, is matter of fact to 'another.'—WHEWELL's *Philosophy of the Inductive Sciences*, vol. i. p. 20.

APPENDIX 1.

COLONEL SIR HENRY JAMES, R.E., *Director of the Ordnance Survey*, ON THE GREAT PYRAMID.

From the Athenæum for November 16, 1867, p. 650.

AND A

CONSIDERATION OF THE EVIDENCE, BY W. PETRIE, ESQ., A CIVIL ENGINEER.[1]

'THE GREAT PYRAMID OF EGYPT.

'ORDNANCE SURVEY OFFICE, SOUTHAMPTON,
November 9, 1867.

'THE publication of the elaborate work on the Great Pyramid of Egypt, by Professor Piazzi Smyth, has led me to an examination of the proportions and dimensions of this Pyramid, in which I have been assisted by Mr. O'Farrell, of the Ordnance Survey. The result of this examination is curious, if only for showing the simple principles on which *the Pyramid was constructed*,[2] and which have been so strangely overlooked by so many ancient and modern authors who have written on this subject.

'First, I find that a pyramid having a square base and a

[1] At p. 189 I have referred to my friend by name, as Mr. Petrie, C.E.; but he writes to correct me in the use of those terminal letters, as he states that he is *not* a member of the 'Institution of Civil Engineers.' His professional tastes are for mechanics and chemistry equally, and they bring him as much engineering employment in connexion with chemical factories, both at home and abroad, as he cares to undertake.—C. P. S.

[2] The *Italics* are now added.—C. P. S.

rise at its corners of nine parts in ten, has the exact proportions of the Great Pyramid.

'Secondly, that the length of one of the sides of the base of the Great Pyramid is precisely 360 *derahs*, or cubits of Egypt.

'This can scarcely be regarded as a mere accidental numerical agreement.

'The derah is a land measure still in use, and is stated by Woolhouse, in his *Weights and Measures of all Nations*, to be 25·488 inches in length; but 25·488 inches × 360 = 764 feet,[1] which is the exact length of one side of the base of the Pyramid, "with the casing-stones," as measured by Colonel Howard Vyse.

'But the side of the square base being 764 feet (9168 inches), the diagonal is 1080 feet, the half of which is 540 feet, from which deducting $\frac{1}{10}$th, = 54, we have 486 feet for the height of the Great Pyramid; and this is the exact height given by Vyse, and in the Diagram which forms the frontispiece to Taylor's work on *The Great Pyramid*. For the construction of the Great Pyramid, the architect, therefore, had only to set up profiles at the four angles, which had a vertical height of 9 feet or 9 cubits to a horizontal length of 10 feet or 10 cubits, for such profiles would be sufficient to guide the masons at every stage of the work from the base to the apex.

'A pyramid with a square base, and a rise at the corners of nine parts in ten, is, therefore, what has been called a π pyramid, its height being equal to the radius of a circle whose circumference is very approximately equal to the length of the four sides of the base.

'In the Great Pyramid these dimensions are 486 × 2 × 3·1416 = 3053·6 feet; 4 × 764 = 3056 feet.

'In the hope of being able to ascertain the true lengths of the ancient Egyptian measures, I have recently, with

[1] Not exactly so; it = 764·64 feet = 9175·68 inches.—C. P. S.

the assistance of Commander Bailey, R.N., of the Topographical Staff, and in presence of the very obliging officers of the British Museum, very carefully measured the double royal cubit of 14 palms which was found at Karnak, and is preserved in the Museum. The length of the digit is 0·739 inches; palm of 4 digits, 2·957 inches; span of 3 palms, 8·871 inches; foot of 4 palms, 11·828 inches; cubit of 6 palms, 17·742 inches; royal cubit of 7 palms, 20·699 inches.

'The length of the royal cubit was inferred by Sir Isaac Newton from the measurements of the King's Chamber and passages of the Great Pyramid, which were taken in 1638-9 by Mr. Greaves, to be so nearly 20·699—(one of his results was 20·700)—that we may feel certain that the interior measurements were set out with a cubit of the exact same length as that of Karnak.

'The inference that this is the true length of the royal cubit is confirmed by the fact that the cubits on the Nilometer at Elephantine, as given by Sir Gardner Wilkinson, and many others, are virtually of the same length as the royal cubit of Karnak. It would seem also that to get the cubit of the Nilometer measured by Le Père=21·289, they divided the six-palm cubit of Karnak, 17·742 inches, into five parts, six of which parts is equal to Le Père's cubit. So that the graduations of this Nilometer are also derived from the Karnak cubit, as well as those of the Elephantine Nilometer.

'I was much struck with the great accuracy of the divisions on the Karnak measuring-rule, which is much greater than we find in general in the divisions of our own measuring-rules; and I therefore think the lengths of its several parts, as I have given them, may now be safely adopted as the true lengths of one set of the ancient measures of the Egyptians.

'But the measures for the base of the Great Pyramid *were set out*[1] on the ground with the derah or cubit of 25·488

[1] The *Italics* added now.—C. P. S.

inches, and Mr. Greaves says he found a cubit in use in Egypt of 21·888, which D'Anville, Grobert, and others found to be as near as possible the length of the cubits of a Nilometer measured by them on the island of Rhoda, at Cairo.

'The ratio of 25·488 to 21·888 is so nearly that of 7 to 6, or 25·488 to 21·847, that we may safely infer that the derah, or land cubit, is also, like the Karnak cubit, divided into seven palms, and that the Nilometer measured by D'Anville and others was graduated from a cubit of six palms of the derah.

'That the derah should have been preserved unaltered from the time of the building of the Pyramid to the present day is not so much to be wondered at when we remember that, on account of the annual inundations of the Nile, a verification of the land measures was annually necessary. HENRY JAMES, *Col. R.E.*'

COPY of Letter sent on 4th December 1867 to the *Athenæum*, and here given because publicity was not accorded by that Journal to any observations on Colonel Sir H. James's Letter published therein.

THE GREAT PYRAMID:

Colonel SIR HENRY JAMES, R.E., *Director of the Ordnance Survey; and Professor* PIAZZI SMYTH, *Astronomer-Royal for Scotland.*

LETTER No. 1 BY WILLIAM PETRIE, ESQ.

The Great Pyramid monument is of surpassing interest if it be regarded merely as the most ancient of existing structures, still in remarkable preservation, favourable to detailed research: it is unsurpassed for vastness with height, also for skill in some respects, and with evidence of sound and deep physical knowledge, though dating from hoar antiquity: it speaks to us distinctly, not merely through the whole historic period of ancient times, but from 1400 years further back.

Unique as a memorial of our race, and as an aid to antiquarian research,—*must* then this structure be abandoned by modern civilisation for purposeless and reckless destruction by tourists, who (chiefly brought by British conveyances) annually obliterate more and more of its instructive features? This question arises irrepressibly, on our writing at the head of this paper the name of the representative of the British Government in its topographical department, honourably known in connexion with science, geologic, antiquarian, etc.

But our present purpose is with the theory of the construction and meaning of that truest Wonder of the World; no response having as yet appeared to the interesting and very suggestive[1] letter of Colonel Sir Henry James in the *Athenæum* of 16th November, after the review of Professor Piazzi Smyth's *Life and Work at the Great Pyramid.* And here, when two heads of scientific departments in Her Majesty's service differ, who shall decide? In our perplexity let us be only too happy, as the saying is, if we can admit them to be men of distinction—without a difference. The absence of difference does helpfully extend through several material elementary theses, whatever divergence there may be on further important points.

Let us note, then, seriously, what theses towards a theory of the Great Pyramid are *mutually recognised* by both parties *and considered settled:* even when regarded from such opposite points of view as by the Astronomer and *savant* with the Scriptures reverently in hand; and by the gallant Director of Military Engineers, surveying *en masse* the whole of the country, with scientific and constructive

1 We say this notwithstanding his light rejection of what we hold the true morphological principles of the Structure, first noticed by the late John Taylor, since then extended by C. Piazzi Smyth, and in confirmation of which we propose to give the shorter portions of the evidence.—W. P.

skill also in plans and profiles of fortification, and an eye for practical simplicity therein; all brought to bear on the field of peaceful antiquarian research. Thus;

A. That two different standard units of measure were used in the Great Pyramid; one for the Base and connected parts, another for most of the ordinary details throughout the Structure. This latter we may briefly speak of as the Chamber-unit, and the former as the Base-unit: to avoid names on which there may be diversity of opinion.

B. That this 'Chamber-unit' was 20·7 British inches, and (always excepting the Base-unit) is the most remarkable of man's units of measure, being the oldest by fourteen centuries, and, there is reason to suspect, vastly older than that. And further,—in this we are more particularly indebted to Sir Henry James, though on other points we are exhibiting the concurrent testimony of the author of *Life and Work*,—this standard was preserved with extreme accuracy down to the times of the example at Karnak, and even onward to our own era. Indeed, the many cubits of comparatively recent, that is historic, times, and in other nations,—cubits of which the varieties in length and subdivision have perplexed metrologists,—were mere derivations from this Chamber-unit; and some, from the yet more remarkable Base-unit. These two,—however contrary in origin and character,—were the exact and pre-eminent standards of unfathomable antiquity.

Sir Henry James has sufficiently proved the above-noticed process, an intrinsically probable one, of derivative variations of metric units, or at least he has well indicated the nature of it, in addition to the changes caused by accidents and bad copying.

C. That the 'Base-unit' was equal to 25 British inches and a fraction. Less than 25½.

D. That the 'Base-unit' was essentially and notably an *earth-measure* in one sense or another, probably in both, be it suggested. The fact itself is here noticed because on one view it has a special significancy, though not within the scope of these theses.

E. That the π angle is true *in fact*; for, it will be allowed to follow from the 9:10 diagonal profile, even by those who decline to regard the converse proposition, that the former was the primary intention, or reason, and the latter a mere inevitable corollary from it.

F. That the Base was originally intended to represent the *days of the year*: for, notwithstanding the material difference between 360 and 365 or 366, it is generally allowed that the artificial 360 divisions of the circle were originally intended to represent the days of the year, and were made 360 either from a misconception of the true length of the year, or from a desire to use, in the practical division of material circles, the divisible number nearest to it, and which, therefore, would still represent the year.

Preliminary to thesis G. We notice that—not only on Taylor's and Piazzi Smyth's theory, but even on that of Colonel Sir Henry James, the features of which we are more especially noticing,—the inference from F is, not only that *the year* is strongly indicated, but, *à fortiori*, self-evidently, the more general idea of A CIRCLE was essentially intended in the original designing of the Base, though put in a square form: an important symbolic reason exists for this, but forms no part of our present subject. We fully expect to be met with the query whether we are to suppose that because such degrees were originally derived from the year, they could never be afterwards applied to totally other purposes; even to the losing sight of their origin, as is the case with the meanings of words? We by no means deny this; but, when a line not circular is composed of such units, and without any manifestation of other sense in it, such composition of the line becomes the most distinct possible token that it has something to do (if not with the originating and concrete idea of a year, yet at least) with the thence derived ideas of *circularity*, which, being abstract and general, would be probably permanent, as in fact it continues to this day: and this circularity, combined with the 51° 51′, etc. angle (no matter by what practical rule set out—whether by Piazzi Smyth's or by Sir H. James's), defines a vertical radius equal to and co-central with the horizontal radii betokened by that circularity.

Thus we have a radius set upright on its parent centre and its own exact circle's length delineated symmetrically around it on the ground, and subdivided in a way that labels it *a circle*. Both the elements of the π construction being thus set forth, each by an independent token,—even if we omit all confirmatory facts, as for brevity's sake, we here do,—the conclusion becomes inevitable, that the π idea was perceived, nay intended, as such, in the original designing of the work: in other words, as follows.

G. That the π construction is true *in theory; i.e.* in the intention of the original design, in addition to the previously mentioned thesis D that it is true *in fact*.

Whatever may be the result of an examination of further questions, they may well be remitted to a subsequent paper. By maintaining, with the Scottish Astronomer Royal, so many positions, from an opposite point of view, Sir Henry James has oppositiously manifested their indisputable character, and has thus aided palæologic truth. In the world of opinion he has advanced our lines of undisputed possession so far, into the unsettled territory of Pyramidology, by setting forth, independently, his view of a *Derah* Base and diagonal profile; involving, as necessary consequences, the above theses. WILLIAM PETRIE.

LETTER NO. 2 BY THE SAME.

All the foregoing foundation-theses are identical with those of J. Taylor's and P. Smyth's theory; and seeing that they result so clearly, though in some cases unintentionally, from Sir Henry James's own view of the case, it seems inconsistent that he should state that view in the manner he does; as if it superseded, or at least disproved, the more complete view, by Piazzi Smyth, of the 'mathematical 'principles on which the Pyramid was constructed.' This seems the less warrantable while Sir Henry James's view depends so much on a single, and as yet unverified, statement (of the precise value and status of the *Derah*) in a small volume in a rudimentary series—for which we own sincere respect, to the extent of its modest profession—without reference to the prior authority from which this datum may have found its way into that work. We take leave to add, that total obscurity and uncertainty rests on the authority and degree of precision of that datum during the very present time in which Sir Henry James is unhesitatingly building thereon no less a superstructure than an original and would-be opposition theory of the Great Pyramid as against what he terms the 'elaborate' work, resulting from the Professor's lengthened and arduous investigations, both literary and physical, on the subject.

But, granting all that Sir H. James assumes to be authenticated concerning the *derah*, how then do the grounds of preference lie? Let us compare categorically some of the respective claims of the supposed *derah* with those of the cubit of 25·025 British inches, as the original base-unit; so far as their respective merits are susceptible of exposition in a few lines. In adjacent opposite columns we will compare, first, their independent existence and importance, and their antiquity, apart from any supposed relation to the Pyramid: next, we will compare their applicability to the Pyramid in respect of accuracy of coincidence, and the nature of the ideas expressed by the resulting number of units in the base and its belongings.

HYPOTHESIS THAT THE BASE UNIT WAS

EQUAL TO 25.488 BRIT. INCHES.	EQUAL TO 25.025 BRIT. INCHES.
FIRSTLY— *The claims of each Standard irrespectively of its relations to the Pyramid.*	
K.—This unit is supposed to be the derah, a measure said to be at present known in Egypt.	K.—This unit is the most probable length of the particular cubit referred to in the real communications of God with his obedient people, as recorded in the Scriptures, in various ages; even that sacred cubit which was used by Moses and the Israelites after leaving Egypt, and by their patriarchal line ages earlier: in the times of structures and events mentioned in terms of the cubit, and which were divinely commanded and recorded. For the evidence of this, see observations on 'Sir Isaac Newton's Dissertations on Cubits,' etc., in *Life and Work*, vol. ii. pp. 341-366, and 454-460, and Appendix, pp. 2-4. There are, besides, indications that this sacred cubit was the original and universal, the Divinely-appointed and natural unit for exact and permanent measurement. It was used for the temple of the true God, and for its contents, in Solomon's time. There are also existing examples, in widely different parts of the world, of anciently inherited standards of measure identical with this,—as nearly as probable errors in our information of the lengths, and in their

BASE-UNIT, 25.488.	BASE-UNIT, 25.025.
	transmission during 2000 or 4000 years, would lead us to expect; and even more accurately. These date back probably to a common origin, as stated.
L.—Its especial use is claimed, by those who assert its reality, to be as a *land-measure*.	L.—It is the natural metric standard for the whole human race; based on the best-defined, most invariable, and simplest dimension of *the earth;* being 10^{-7} × the polar radius; *i.e.*, its precise ten-millionth; free from the objections attaching to other earth-standards since proposed by human science. It is, likewise, most remarkably, just the four-thousand-millionth part of the earth's mean daily march in its orbit, reckoning the *true* rotation or sidereal day, not the apparent day.
M.—From its existence as a cubit of seven palms (says Sir Henry James), may have been derived the six-palm cubit of 21·847 inches, such as that in Egypt found by Greaves=21·888 inches.	M.—From its existence as a cubit of seven palms, may have been derived the six-palm cubit, 21·45 inches; the probable origin of a particular class of cubit examples, such as those in Egypt, found by Le Père, Sir G. Wilkinson, and Greaves, 21·29 and 21·4 and 21·88, whether they were in course of time differently subdivided or not, and slightly increased or decreased in repeated copying in later days of inferior artistic accuracy; and by wear and accident, and by attempted corrections for such changes.

BASE-UNIT, 25.488. | **BASE-UNIT, 25.025.**

SECONDLY—

The claims of each Standard by its relation to the Pyramid.

N.—It subdivides the base-side into 360. This was the mode of division for the circle, etc., invented at Babylon, about 1500 years after the Great Pyramid was built; and our best authorities consider that it was not in use previously. This subdivision, therefore, of the base, needs collateral proof that this system of subdivision was in use in the Pyramid, or in those times: such proof, for instance, as we have in many parts of the Pyramid, that five is the number *par excellence* characteristic of it (see *Life and Work*), and that other numbers were especially used in certain places in it, on a fixed principle; whereas 360 does not seem to have sufficient probability on this ground.

The best that can be said of the possible connexion of such existing Egyptian measures with the base of the Great Pyramid, is, that if the existing Egyptian *gasab* (for this will suit the case better than its fourth part, the derah), be descended from far beyond pre-historic antiquity, or even probably so, and its ancient length were (101·839) = ·113 of an inch, or about $\frac{1}{1000}$ part shorter than our present information assigns as its existing length, it would then become probable that this measure was derived

N.—It subdivides the base-side 9,165·6 Brit. inches (see p. 408), into the days of the year, reckoning the true or sidereal rotations of the earth in the course of its orbit, 366·257. Moreover, the subdivision of this cubit into 25 parts or Pyramid inches,—for in *Life and Work* this is explained to be in harmony with the principles of the Pyramid arrangements,—shows each day as being exactly 100 Pyramid inches, or a fourfold cubit, around the entire circuit of the base.

This coincidence, being significant, cosmically, is on that account more probable than any other suggested division. But this probability is highly increased by the fact that this same division or unit of the base (25·025) establishes several other dimensions with unequalled simplicity.

This unit makes the co-efficient of the Pyramid's height or radius to be expressed with remarkable neatness, namely $\frac{7}{3}\times 100$. The digits 7 and 3 being marked in the construction of the ascent (or *height* passage) to the so-called King's Chamber, besides being remarkably connected (as shown in *Life and Work*, Postscript, vol. iii.), with the decimal expression of the π ratio on which this

from the Pyramid base, divided by 90; because 9 and 10 in combination are numbers connected with the Pyramid in at least two other ways, the diagonal profile and the relation of its radius to the solar distance, and therefore, by a *moderate* probability, these numbers may have been subsequently taken to divide the base for a new unit;—while the base would have been formed, nevertheless, on the 25·025 unit, for, on that unit it represents the earth's true rotations in the complete orbit. But, it may be asked, if *both* these units are probably or possibly true, why assume the 25·025 unit to be the primary intention, and the 101·839 or its quarter (a length approximating, though not exactly, to the derah) to be the unit subsequently derived from the base? why not *vice versâ*, as Sir H. James's theory would suggest? The reason is, 1st, that the 25·025 unit is more certainly a palæologic fact than the 101·839 unit; it is so on grounds here in course of being shown; and, 2dly, if *both* be palæologically real, then the round subdivision of the base (into 90 parts, whether by the author of the design, or probably by others without authority) to obtain a convenient unit of measure for a particular use, as land measuring, is a process greatly more probable to have

height depends. All this is suggestive of an intended connexion rather than a mere coincidence.

This unit makes also remarkable cosmical expressions—

Earth's mean sidereal or correct day's march,	4×10^{9}
Sun's mean distance,	$\frac{1}{4} \times 10^{11}$
Earth's polar radius,	10^{7}

These facts are of precisely the same class, nay more, of the same order, and even of the same family group with the days in the year already noticed: and we do not reasonably expect to find such a connexion among coincidences of a purely fortuitous origin.

BASE-UNIT, 25.483.

been performed than the converse; that is, than commencing with an arbitrary, or at least an *uncosmical* standard (the gasab or the derah), and contriving—from a round multiple of it (the base length), a re-subdivision that should correctly produce, at one operation, exact tokens of two independent cosmical facts,—the year, and the earth-radius.

This idea of the gasab, as a unit of measure originated by the Pyramid base after its completion, is merely suggested to one's mind as obviating the want of probability in the converse idea of Sir Henry James, that the length of the Pyramid base was itself originated by the derah. But until more is known of the antiquity or origin of the gasab (or of its quarter, the derah), and of the probable limits of error in its assumed length, its derivation from the Pyramid base must remain a mere suggestion, though somewhat more probable than that of the derah, and gasab; or than either of these having originated the base-length.

O. The relation of the derah to the other parts of the structure, is *nil.*

BASE-UNIT, 25.025.

O. The relation of the sacred cubit to other parts of the structure is confirmatory of its having been the true standard employed: though this is certainly the least decisive branch of the evidence in our present very imperfect

BASE-UNIT, 25·025.

state of knowledge of the parts not yet uncovered, and of parts destroyed because prematurely uncovered.

Some of the four base-corner stones (perhaps all, if we knew all their dimensions) show the precise double or quadruple of the sacred cubit in their hidden diagonal dimensions,—their diagonals either of surface or of solidity. We might add that the probable proportions of some of these stones, in terms of the sacred cubit, are such as would be of much interest were our information less liable to uncertainty.

Again, there is strong reason to believe that the horizontal thickness of the casing was on the mean 100 Pyramid inches, or 4 sacred cubits.

Again, in the interior there is apparently a use of the sacred cubit, or its inches, in particular dimensions in the so-called King's and Queen's Chambers.

BASE-UNIT, 25·488.

P. The imperfectness of the coincidence of the asserted derah-multiplicate with the Pyramid base-length, is such as to show an excess of 10·2 inches beyond the base-length, according to the most judicious mean of the only two trustworthy measurements which we possess: those of the French *savants* and of Colonel Howard Vyse. Or even if we cook the result by selecting the

BASE-UNIT, 25·025.

P. The extreme precision of the coincidence of the 25·025 inch standard, or sacred cubit, (with the division of the base-length by the rotations of the earth in the course of its orbit) expresses that it was intentional and not fortuitous, as strongly as coincidence can express this with our present data: for, the coincidence is far more close, indeed, than the small extent of uncer-

BASE-UNIT, 25·488.

most favourable measurement, the want of coincidence is yet 7·7 inches. This is very unlikely to be due to errors of measurement, for they must be in the contrary direction, considering the causes of error in the termini of the base, and in the process of measuring, and the results of ruder measurements of others when the rubbish intervened. But it might be accounted for by the probable alteration of the unit of measure in the lapse of 4000 years. But while we thus hold that the want of coincidence in Sir H. James's derah-base is readily accountable, it is sufficient to prohibit the term *precise*, which Sir Henry James applies to this moderate coincidence, since he applies the term *approximate* only to the much more close coincidence of the π construction advocated by others.

BASE-UNIT, 25·025.

tainty in the true mean between the only two reliable measurements of the original base. That of the French, 9163·44 British inches, and of Howard Vyse, 9168, or 9167·50 after making a small allowance for wear since the first uncovering of the sockets, and for the fact that Vyse did not notice, and probably did not allow for the existence of an intervening film of cement, as the French: the most probable truth will be (9163·44 + 9167·50) ÷ 2 = 9165·47. Now the number of the 25·025 inch base-units in this, agrees with the rotations of the earth in the course of its orbit exactly to a hundred thousandth of the whole quantity: indeed much more exactly than our uncertainty of the length of the base.

What a contrast to Sir Henry James's coincidence of the derah-multiplicate which he calls exact, although it shows about 100 times the error!

Such are the comparative merits of the derah (25·488) advocated by Colonel Sir Henry James, as a base-unit, as compared with the sacred cubit (25·025) advocated by Professor Piazzi Smyth. The comparison is not in favour of the derah, in any aspect of the case: and yet—excepting the sacred cubit—the derah is really the best out of the multitude of units of measure that have been tried, by persevering industry, with the purpose of finding some unit that shall show an intelligible quotient on dividing the base, and at the same time shall have at least some shadow of evidence of having had an independent existence as a standard of measurement among men in those days. How strikingly, then, does the above-shown superabundant fulfilment of these conditions, by the sacred cubit, contrast with the failure of all other proposed standards!

WILLIAM PETRIE.

LETTER No. 3, by the SAME.

Review of the subject, and general considerations.—Colonel Sir Henry James states that the principles on which the Great Pyramid was constructed, were simple, and *have been strangely overlooked* by ancient and modern authors, and that this is his conclusion, arrived at by an examination consequent on the publication of 'the elaborate work' of Professor Smyth.

But it is important to notice, 1st, that the Karnak cubit, which he brings prominently into the question, perfectly confirms much that Piazzi Smyth has written, while it gives no additional information, and opposes nothing in that author's work, although the tone of Sir Henry James's letter would imply the contrary. And, 2dly, that the 'principles' of construction put forth by Sir Henry James do not militate against the *facts* on which Professor Smyth has rested his exposition of the meaning, intention, and authorship of that structure—the great ends for which the facts are really valuable to human intellect; though this, being theory, may be slighted as such, by many. Indeed the facts are common to both; and the whole remaining body of the facts and inferences, concerning the interior of the structure, are not included at all in Sir Henry James's principles of construction: these are meagre enough as principles, relating solely to *the mathematical rules*—irrespectively of their reasons or their meaning—by which *the exterior form only* of the Great Pyramid was constructed. On this point, the opposite rules enunciated by each party, agree with what they assume to be the facts, for the present, and until the facts of dimension and angle shall be, if ever, more accurately—and perfectly impartially—ascertained; chiefly from the casing and corner stones at present built into certain public works at Cairo.

Meanwhile, there is another test by which to decide the matter; let us ask,—is the one set of supposed mathematical rules of construction, a constituent part of a harmonious and well-united system of distinct signs and tokens, embracing a much larger mass of facts (or coincidences, as some may be pleased to name them) in that structure;—and is the other suggested set of mathematical rules, isolated from the facts observed in other features of the structure? or, at least, not indicated by a similar connexion with them. Judged by this test, the evidence is wholly in favour of Professor Smyth's view, as is shown in detail in the third volume of his last work. To the writer's mind this is greatly confirmed by a further tracing out of the same clue; but this being as yet unpublished, he can only assert such additional testimony as being his own impression.

On the other hand, Sir Henry James appeals to the superior *simplicity* of his rules of construction: but this superiority is not found in their physical, nor even mathematical, simplicity—it is only arithmetical. Thus, a certain *radius* for height, and its *circumference* for base length, is quite as simple or plain a rule, regarded mathematically or physically, as Sir Henry James's alternative of a *triangle* with two of its sides forming a *right angle*, and of *unequal length* in a certain ratio, though that ratio be not arithmetically complex, and with *some additional* process for defining the base length. If the angle of the diagonal profile had been a plain half right angle, 45°: or 1 vertical in 1 horizontal, there would have been a tolerably good balance of probability between that and any alternative angle as simply defined, and as coincident as the existing π angle of John Taylor and Piazzi Smyth. But the diagonal profile is not simply 45°, it is 41° 59′ 50″ etc., or approximately 9 : 10 ;[1] and as the π theory and this are mathematically inseparable, or each one is a necessary consequence of the other, why should the one which is the less simple, less cosmical, and less extrinsically supported by unity of design, etc., be assumed to have been the original intention? Again, the division of the year into its days,—whether solar, *i.e.* apparent, or sidereal, *i.e.* real—is, mathematically and physically, as plain a fact as the purely artificial division of a circle into 360 parts.

In each of these cases (the profile and the base-length) we have to choose between a most plain fact, abundantly illustrated and deeply rooted in cosmical nature, and an artificial *idea*, recommended chiefly, if not solely, by the brevity of its mere arithmetic, in man's mode of mentally grasping ratios. Hence, in the argument of the probability that simple, natural, and universal illustrations formed the basis of the design of the Great Pyramid, we cannot recognise any preponderance in favour of Sir Henry James's view; while on the previously noticed ground of unity of design and meaning, the evidence is wholly on the side of Taylor's and Smyth's view, as already stated. Although the 9:10 profile *may* have been given to the workmen as a guide, in laying the rough interior blocks, yet the question is, not this but—the principle or *intention* of the design.

The only way of making the probabilities to be against the π theory, as compared with the 9:10 profile theory, is by a series of untenable *suppositions*,—as, that the originator of its design was of low mathematical and constructive attainments, so that he did not even *perceive* the remarkable

[1] We say 'approximately,' because Sir H. James applies this term to the π angle, and their claims to accuracy are equal.—W. P.

π proportion which he hit upon; also that he selected the diagonal profile with its extremely close agreement (exact to about a 5000th part) with the π angle, by mere accident: and similarly as regards the other coincidences (nearly as close, or for aught we yet know, absolutely accurate), the relation to the sun's ray and earth's axis, and its rotation, etc., and their manifest analogic relation to the π construction, that all these and their relations to one another are purely accidental and unintentional. Otherwise there is not a *primâ facie* case in favour of the diagonal profile theory; and, that the structure was on the contrary designed by transcendent ability and science, is amply proved by a combination of evidence, well set forth in *Life and Work.* We say so, even though it be proved that weak points have been included in that provisional collection of various lines of evidence. Now, be it remembered that it is useless to theorize on an alternating basis; admitting, in presence of some combinations of evidence, consummate skill and knowledge in the designer, and then, on another part of the subject, *preferring* one theory to another on the ground that it does not necessitate the hypothesis of that admitted skill and knowledge,—notwithstanding that the theory thus preferred be less commended by unity of design. Whichever theory has been adopted, let it be plainly professed and adhered to throughout the *whole* subject, in its interior, astronomy and exterior, its symbolic and constructive branches; the theory will then show its worth.

Amid an apparently gratuitous divergence of opinion, it is interesting to notice, as in Letter No. 1, how some of the more feebly self-evident, but vitally important theses, on which Professor Smyth's views rest, are confirmed by Sir H. James's chosen data. The relations of the latter to the former are simply, that Colonel Sir Henry James has the credit of calling attention to a reported existing Egyptian measure, which is so similar to the remarkable ancient and still existing[1] unit of Taylor's and Smyth's theory, as to be confirmatory, and suggestive of an original identity; and to establish the same intention of circularity, etc., in the base,—as in theses A to G. Sir Henry James has also the credit of making (not altogether an original, but certainly) an independent discovery, and of being the first to publish —that a Pyramid with a 9:10 diagonal profile is approximately the same mathematical form as the π relation of base and height;—an interesting contribution to the π theory.

WILLIAM PETRIE.

[1] See, for instance, the *guz* (25·025 Brit. inches, within probable errors of measurement), the measure of the Arabs,—that branch of the family of Abraham (that great chosen Patriarch of the national custodians of the Scriptures of the living God and their sacred cubit), that branch whose traditions, measures, language, customs, etc., have been less disturbed by conquest, and by voluntary changes of national habits, or caprices of fashion, through the lapse of ages, than those of any other nation.—W. P.

APPENDIX 2.

DISCOURSES AT ROYAL SOCIETY OF EDINBURGH ON GREAT PYRAMID SUBJECTS;

IN DECEMBER 1867, AND JANUARY AND APRIL 1868,

AS REPORTED IN THE DAILY PAPERS.

WITH REFLECTIONS THEREON.

Meeting of Royal Society of Edinburgh, on December 16, 1867, *as reported in the* EDINBURGH EVENING COURANT *for December* 23, 1867.

'ANTIQUITY OF INTELLECTUAL MAN.

'ON Monday night (December 16) the second ordinary meeting of the Royal Society was held in the Royal Institution. Professor Lyon Playfair, C.B., vice-president, in the chair.

'Professor C. PIAZZI SMYTH delivered an address on "The Antiquity of Intellectual Man, from a Practical and Astronomical Point of View." In his address, which was illustrated by numerous diagrams, the Professor described the leading features of the Pyramids of Egypt, and noticed the remains of proto-Chaldæan buildings, and the earliest remains of buildings found in the valley of the Euphrates, which indicated the existence there of several large cities, which must have flourished long before Babylon was ever heard of. After referring to the remains of Babylon, he proceeded to point out that all these architectural remains, as well as those of the Medes, Persians, Phœnicians, Greeks, and Romans, were all of less antiquity than the Great Pyramid of Egypt, which he dated at 2000 years before Christ. He referred to a statement, made by Sir James Simpson in his recent work on "Ancient British Sculpturings," as to there being "rude but magnificent pyramids" in Ireland and Brittany, maintaining that these Irish remains were exactly opposite to the Egyptian Pyramids in shape; and that they had no place among Pyramids, whether tested by geometry or astronomy, while their mechanical construction was something so wretched, barbarous, and savage-like, that their present tumble-down condition would seem to be explainable by the passage over them of not a third of the number of years during which the Great Pyramid of Egypt had witnessed the rise and fall of nations. The Professor next disputed the accuracy of a statement in the same work of Sir James Simpson's, to the effect that, from the very earliest periods, circles, single or double, and spirals, had formed perhaps the most common fundamental type of lapidary decoration; and adduced the testimony of a gentleman in England, to the effect that the circles referred to were the last efforts of nations before lapsing into barbarism. He also noticed a statement made in a paper read before the last meeting of the British Association, to the effect that there were writers known to have written in Egypt so far back as 9800 B.C.; and contended that the authority on which that statement was made was utterly unreliable, while the statement itself was opposed to well-known facts as to the deposits found in the valley of the Nile. Professor Smyth proceeded to refer to the latest theories which had been propounded as to the principles on which the Great Pyramid was constructed, and endeavoured to show, from various measurements of the Pyramid, that its designer must have been acquainted with many of the discoveries of modern astronomy, such as the distance of the earth from the sun, and the precession of the equinoxes. He maintained that the Great Pyramid showed, by a series of scientific proofs, that miraculous aid must, in some manner or other, have been given to its designer, to have enabled him to accomplish so wonderful a work at such a period of the history of the earth.

'Sir JAMES SIMPSON, on being called upon by the President, said he regretted that his friend Professor Smyth had not given him the slightest notice of his intended attack.[1] But in reply to Professor Smyth's criticisms on his work, he argued at some length that the Professor was greatly mistaken in his ideas of the ages of different schools of architecture. For example, he had spoken of Greek architectural remains dating back only to the seventh century B.C.[2] But every one in the slightest degree acquainted with Greek archæology, knew that there were far older architectural remains in Greece, as in the Cyclopean walls and Lion's Gate at Mycenæ, which still stand wonderfully entire. The curious building known as the treasury or tomb of Atreus, at Mycenæ, was reckoned by all the best Hellenic archæologists as being several centuries at least older than what Professor Smyth gives as the oldest specimens of Greek architecture. But this tomb of Atreus was more interesting in another point, alluded to by Professor Smyth. Professor Smyth challenged—what all architects

[1] There was no 'attack' made or intended; nor was Sir J. Y. Simpson mentioned by name. Two passages in a published work by him, expressing opposite opinions to those of the author, were *reviewed*; as were many other persons' opinions upon both that one section, and the numerous other sections of which the address consisted. And its author would not have been held excusable in scientific society, had he omitted to notice published opinions of great men unfavourable to his views.—C. P. S.

[2] In this sentence, and that which follows, Sir J. Y. Simpson does not notice that it was only the Hellenic Greek architecture whose rise was placed by the author in 700 B.C.; while in another column of his Chart (see the Frontispiece) the author had given the Pelasgic architecture of Greece an antiquity of 1800 B.C., agreeably with Mr. Fergusson's teaching.—C. P. S.

certainly allowed—that circles and spirals are among the earliest and fundamental forms of stone decoration; and he held that circles and spirals marked the period of degradation, not the period of commencement in lapidary ornamentation. But the pilaster at the door of Atreus' tomb—probably eleven or twelve centuries B.C.,—was elaborately ornamented with circles and spirals, and yet it is among the oldest (not the youngest) and degraded specimens of Greek art. The temples of Thebes, Assyria, showed abundance of decorations originating in spirals and circles, and in pieces of them—are all geometric. Any of the most common books on ancient architecture would convince Professor Smyth of this. He had no doubt whatever that the Pyramids of Egypt, as of Ireland and Brittany, were all tombs. Professor Smyth tried to speak disparagingly of those of Ireland and Brittany, because they were less in size;[1] but a planet was still a planet, whether enormously large or very small. The sight of the old Irish mausolea, scattered for miles along the banks of the Boyne, was one of the most remarkable archæological sights in the United Kingdom—tall and large as little hills, like those of Newgrange, and showing in their interior large cyclopic chambers, the huge component stones of which were sometimes highly decorated. If Professor Smyth would visit and measure them, he would perhaps find them as interesting and astonishing as others he had visited, or more so. Those of Brittany—many of them decorated interiorly—were perhaps older than those of Egypt, for they contained implements and weapons only of stone—not of bronze or iron. They were reared in the stone age—the Egyptian tombs were reared after metallic tools were in-

[1] This statement is erroneous. Professor Smyth disputed the right of the Irish tumuli to be called 'pyramids,' solely on the question of shape; and he not only exhibited a model of a good pyramid, so small as 12 inches high,—but also gave the geometrical definition as applicable to any size, large or small.—C. P. S.

vented. But if not content with the age of these enormous western tombs or pyramids, we had, besides cromlechs and megalithic circles, other remains of intellectual man in Europe undoubtedly older than the oldest Egyptian Pyramid. He then alluded to the carving upon a piece of ivory of a figure of the mammoth—moving or running—found by M. Lartet and Dr. Falconer in one of the old inhabited caves of France. That accurate piece of drawing or sculpture must undoubtedly have been executed by a man that saw the mammoth with his own eyes. Yet surely Professor Smyth would admit that the date at which this "intellectual" man lived in France was far, far older than the age of the oldest pyramid. In the most ancient grave of man yet discovered, at Aurignac, a piece of sculpture was found, consisting of a mammal tooth, shaped into the form of a bird's head. Yet we know that those buried in this grave were contemporaneous in France with the mammoth, rhinoceros, Irish elk, etc. In reference to the question of the astronomical questions connected with the old Egyptian Pyramids he was no judge. But he understood, though all the measurements were near, none were exactly what the theory wanted. But things mathematical, to be mathematical, should be correct entirely, and not merely nearly so. He concluded by remarking that though Professor Smyth's papers were extremely clever, he did not believe one word about the Pyramids having been built on astronomical principles. He had no faith whatever in what was called the astronomical size and measurement of the Pyramids: they might as well tell him that the Register Office and the College were arranged on a similar principle. It was a spurious archæology and astronomy that fed on such fancies and romances.

'The company then adjourned to the next room, and partook of tea before separating.'

The last Speech of the above Meeting, as given the next morning, or on December 17, 1867, *in the* SCOTSMAN.

'Sir JAMES SIMPSON, in reply to Professor Smyth's criticisms on his work, argued at some length that the Professor was greatly mistaken in his estimate of the antiquity of the architectural remains of Greece, and gave it as his opinion that the Pyramids of Egypt were nothing but tombs, and that there were tombs in Europe older than the Egyptian Pyramids—a fact which he maintained to be proved by the discovery that they were the tombs of persons who lived before metal weapons were known in Europe, while the Egyptian Pyramids were built after metal weapons began to be known. He asserted that the statements in his work which had been called in question by Professor Smyth were accurate; and, in proof of the fact that there were remains in Europe more ancient than the Egyptian Pyramids, he referred to pieces of sculpture which had been found in graves, which showed that the sculptor must have seen the mammoth. Sir James concluded by remarking that though Professor Smyth's papers were extremely clever, he did not believe one word about the Pyramids having been built on astronomical principles. He had no faith whatever in what was called the orientation of the Pyramids; they might as well tell him that the Register Office and the College were arranged on a similar principle. The company then adjourned to the next room, and partook of tea before separating.'

Meeting of January 6, 1868, *as reported in the* SCOTSMAN *of January* 7.

'ROYAL SOCIETY.—The third ordinary meeting of the Royal Society was held last night in the Royal Institution.

Dr. Christison, vice-president, occupied the chair. The Chairman delivered the Keith medal to Professor Piazzi Smyth for his communications during the last biennial period, entitled "Notices of Recent Measurements at the Great Pyramid." The Committee, in the minute recommending the Council to make the award, recorded their opinion "that the energy, self-sacrifice, and skill displayed by Professor Smyth in this vast series of measurements fully entitle him to the highest testimony which the Council has in its power to offer." The Council, Professor Christison remarked, offered no opinion of the theory propounded by Professor Smyth, but they were fully convinced that this was a most valuable series of measurements, on which all future theories must be based. Professor Piazzi Smyth, in returning thanks for the honour conferred on him, expressed regret that the stones of the ancient tombs of the Egyptians, and even of the Great Pyramid itself, were being carried away for the construction of public and private buildings, so that there was reason to fear that, after being so remarkably preserved for forty centuries, they would yet pass away from the earth; and, under these circumstances, he hoped that the notice which the Royal Society had taken of his labours, would direct the attention of other investigators to this subject. Dr. A. Crum Brown and Dr. J. R. Fraser read communications "On the Modifications produced by Direct Chemical Addition upon the Action of Certain Poisons." Professor Kelland submitted a paper by Mr. John Scott, "On the Burning Mirrors of Archimedes, and on the Concentration of Light produced by Reflectors." The object of the paper was to show how certain defects in the experiments by Buffon might be remedied, so as to form a strong presumption that the accounts of the Archimedian mirrors were authentic. Professor Kelland also read a communication by A. D. Wackerbarth, of Upsala, on Professor Piazzi

Smyth's former account of the Great Pyramid of Gizeh. The writer was of opinion that the great Pyramids might have been the standard of the ancient Egyptian weight and measure, but he did not believe that it was designed in accordance with the theories that had been propounded by Professor Smyth. After reading the paper, Professor Kelland made some observations as to the value of the measurements which had been made by Professor Smyth, and showed how they demonstrated the truth of Pliny's statement of the dimensions of the Great Pyramid. Sir James Y. Simpson replied to Professor Piazzi Smyth's criticism of his work on "Ancient Sculpturings" at last meeting of the Society. He did so with the utmost friendship for Professor Smyth, whose measurements were most exact, but whose logic was most wretched. In speaking of the circles and spirals found on the stone monuments in this country and elsewhere, he (Sir James) suggested that they were sepulchral and ornamental. He had since last meeting shown the passage referred to to Mr. Bryce and Mr. Kinnear, and they quite agreed with him in the opinion he expressed on the subject. (Sir James said that) Professor Smyth admitted that they might have been so used, but (that) he (Prof. Smyth) said that he (Sir James) had made a slip in saying that these circles and spirals had been used from the earliest periods. Sir James quoted from various authors, amongst others Professor Smyth himself, to show that circles were used on these monuments for purposes of adornment from the very earliest times.[1] After transacting some private business the members and their friends had tea served in the anteroom.'

[1] No reply was made to this discourse: which was irregular, had not been allowed by the Council, nor announced in the billet, did not refer to the advertised business of the evening, and occupied so long as to prevent the last of the papers on the programme being read.—C. P. S.

Meeting of January 20, 1868, *as reported in the* DAILY REVIEW *of January* 22.

'ROYAL SOCIETY OF EDINBURGH.

'SIR J. Y. SIMPSON AND PROFESSOR PIAZZI SMYTH ON PYRAMIDAL STRUCTURES.

'On Monday night, an address (which we briefly noticed yesterday) was delivered to the Royal Society, by Professor Sir James Y. Simpson, on "Pyramidal Structures in Egypt and Elsewhere, and the Objects of their Creation." There was an unusually large attendance of members.

'Sir JAMES Y. SIMPSON commenced by stating that the probable derivation of the word *pyramid*, as suggested by the distinguished Egyptologist, Mr. Birch, was from two Coptic words signifying the "king's sepulchre," and he then pointed out, that in various parts of the world, very large sepulchral conical hills existed, made sometimes of earth and sometimes of stone, containing within them relatively small chambers for the dead, constructed of enormous stones, and galleries often concealed, leading into these chambers. He cited various authors using the word "pyramid" in this sense, and particularly Sir William Wilde, who, in his travels in the East, when visiting some of the Egyptian Pyramids, tells how their dome-shaped interior chambers reminded him of the Pyramids in Ireland at Newgrange, Louth, etc., and again, in his work on the Boyne, the same distinguished archæologist discusses, at greater length, the analogy between the Pyramids of Egypt and the Pyramidal mounds in the west of Europe. All authors, from the father of history downwards, have generally agreed in describing the Pyramids as magnificent royal sepulchres; and the sarcophagi of dead princes have been found in many—if not indeed in all—of them, when first opened. The Pyramidal sepulchral mounds on the banks of the Boyne were opened and rifled in the ninth century;

and those of the Pyramids of Gizeh were broken into and harried in the same century by the inquiring Arabian Caliph El Mamoon. By some, the largest of the Pyramids at Gizeh has been maintained, and particularly of late by Gabb, Jomard, Taylor, and Professor Smyth, to be a marvellous metrological monument, built some forty centuries ago, and containing within it, and in its structure, standards for men and nations, for all times, of measure of length, capacity, weight, etc. Messrs. Taylor and Smyth further hold that this Great Pyramid, as it is termed, was so built for this purpose of mensuration under Divine inspiration; the standards being, under superhuman origination and guidance, made and preserved in it to the present day. It was the largest of the many Pyramids of Egypt, but apparently not the first, and had been calculated by Mr. Furlong as a structure which in the East would cost about £1,000,000. Over India, and the East generally, enormous sums had often been expended on royal sepulchres for the dead; and the Taj Mahal of Agra, built by the Shah Jehan for his favourite Queen, cost perhaps triple the money. Mr. Taylor ascribes to Noah the original idea of the metrological structure of the Great Pyramid. "We are told," says he, "that Noah was a 'preacher of righteousness,' but nothing could more illustrate this character of a 'preacher of righteousness' after the flood, than that he should be the first to establish a system of weights and measures for the use of all mankind, based upon the measure of the earth." Professor Smyth, computing by another chronology, rejects the presence of Noah, and makes a shepherd—Philition—the presiding genius, holding him to be a Cushite skilled in building, in the same way as Mr. Keane, in the last-published book on Irish archæology, strangely holds the round towers, etc., of that country to be all also Cushite in their origin and masonry. A granite coffer, stone box, or sarcophagus standing in one of the interior cells of the Pyramid called the King's Chamber, is held by Messrs. Taylor and Smyth to have been hewn out and placed there as a measure of capacity for the world, and that our Hebrew measures on the one hand, and our Anglo-Saxon on the other, were all derived from the measurements of this granite vessel. The lecturer read various extracts from the works of these authors, showing that in their belief the great object, end, and aim for which the whole Pyramid was created, was the preservation of this coffer as a standard of measures. He read the published accounts, averring that it represented without and within a rectangular figure of mathematical form, perfectly level, and highly polished. But though thus imagined to be a great standard of measure, Professor Smyth had cited the measurements of it, made and published by twenty-five different observers, several of whom had gone about the matter with great mathematical accuracy; but all these twenty-five, as he owned, varied from each other in their accounts of this standard in every element of length, breadth, and depth, both inside and outside. Professor Smyth had measured it himself, as a twenty-sixth measurer, and again his measure varied from every other. Yet surely a measure of capacity should be measureable, but this seemed immeasureable—at least, to twenty-six different measurers. "Still," says Professor Smyth, "this affair of the coffer's precise size is the question of questions." He holds that theoretically its capacity ought to be 71,250 cubic inches, yet he himself found it some 90 cubic inches less than this; while Professor Greaves, one of the most accurate measurers of all, found it 210 cubic inches below, and the French academicians who accompanied the Napoleonic expedition to Egypt, found it about 6000 cubic inches above, the theoretical size requisite to make it a measure for a chomer or chalder—namely 71,250 cubic inches. Thirty years ago it was pointed out, and indeed delineated by Colonel

Howard Vyse, as not well polished, and much broken at the edges. Professor Smyth, in his late travels to Egypt, states that he found every possible line and edge of it chipped away with large chips, and one corner broken off to nearly a third of the whole. But worse facts than these are revealed of it by Professor Smyth as a measure of capacity. His published measurements show that it is not a vessel, as alleged, of pure mathematical form; for externally it is in length an inch greater on one side than another; in breadth half-an-inch broader at one point than at some other points; in thickness of sides it varies in some points about a quarter of an inch; and its bottom at one point is nearly a whole inch thicker than it is at some other parts. Internally there is a convergence at the bottom towards the centre; in length and breadth it differs about half-an-inch at one point from other points, and thus its interior seems really of a form immeasureable in a correct way by mere lineal measurement—the only measure yet attempted. If it were an object of the slightest moment, perhaps liquid measurements would be more successful. Worse details still remain in relation to the coffer as a great standard measure of capacity. Its western side is lower down in its whole extent to nearly an inch and three-quarters; and ledges are cut round the interior of the other sides at the same height. Should we measure on this western side from this ledge, or from the imaginary brim? Besides, there are three holes cut on the top of its lower western side, as in all the other Egyptian sarcophagi, and used with the ledge to admit and lock the lid of such stone chests. In other words, it is the ledge and apparatus pertaining to the Egyptian sarcophagi. When about a thousand years ago the Caliph Al Mamoon tunnelled into the interior of the Pyramid, he discovered by the accidental falling of a granite portcullis, the passage to the King's Chamber, shut up from the building of the Pyramid, to that time. The Arabian historian Ibn Alhokm states that on this occasion they found in the Pyramid, "towards the top of a chamber (now the so-called King's Chamber) within a hollow stone (or coffer) in which there was a statue like a man, and within it a man upon whom was a breastplate of gold set with jewels; upon this breastplate was a sword of inestimable price, and at his head a carbuncle of the bigness of an egg, shining like the light of the day, and upon him were characters writ with a pen, which no man understood,"—a description stating down to the so-called "statue," or painted wrapper, or cartonage, and the hieroglyphics upon the sere cloth, the arrangements now well known to belong to the higher class of the Egyptian mummies. In short, the marvellous metrological coffer, mathematically formed under alleged Divine inspiration as a measure of capacity (and, according to M. Jomard, of length also), for all men and all nations, for all time, and particularly for these latter profane times, is in truth, proceeded Sir James, nothing more than an old dilapidated stone coffin. The standard in the Great Pyramid, according to Professor Smyth, for lineal measurements, is the length of the base-line or lines of the Pyramid. It is pretended that in this base-line there has been found a new mythical inch—one-thousandth of an inch larger than the British standard inch; and Professor Smyth has attempted to show that the status of the nations of Europe in the general and moral world, may be measured in accordance with their deviation or conformity to this mythical Pyramidal standard. The engineers and mathematicians of different countries have repeatedly measured a meridian arc of a league of sixty miles, in order to employ it as a standard for lineal divisions. As part of their standard, they measure off sixty miles of the surface of a kingdom with almost perfect mathematical exactitude. Professor Smyth, continued Sir James, holds that one side of the Pyramid has been laid down by Divine authority, as such a guiding standard measure. The sides of the Pyramid have been measured by many measurers. But, as in the case of the cubic capacity of the coffer, none of these two measures agree with each other, but vary from some 8050 inches to 9050 and upwards. At the time Professor Smyth was living at the Pyramid, Mr. Inglis of Glasgow visited it, and laid bare for the first time, the four corner sockets. But Mr. Inglis's measurements not only differed immediately from all the others made before him, but he made the four sides differ from each other; one of them being longer than the other three. Strangely, Professor Smyth, though in Egypt for the purpose of measuring the Pyramid, and holding that its base-line ought to be our standard of measure, and that the base-line could only be accurately ascertained by measuring from socket to socket, never made that measurement himself; but he lays down as the measurement of one side the figure 9142 inches (which no one ever found it to be), because that figure will be the result of multiplying the number of the days of the year, by what he calls the Pyramidal cubit, made up of twenty-five mythical Pyramidal inches, and which Pyramidal cubit he further holds to be the same as the "sacred cubit" of the Jews, being given by Divine inspiration. In proof of the length of the "sacred cubit," Professor Smyth cites the essay on the subject by Sir Isaac Newton, who long ago showed, from Professor Greaves' measurements, that the working cubit of the masons in the Great Pyramid, as evidenced in the interior of the building, was the old Memphis cubit. Sir Isaac Newton twice lays it down that the sacred cubit is in length 24·82. But for Pyramidal purposes, Professor Smyth thinks it should be 25·07. In obtaining this end, Professor Smyth makes a wrong summation of Sir Isaac Newton's figures, calculating nine means, which he gathers out of Sir Isaac's essay, as amounting on a mean to 25·07, when they actually amount to 25·29, while he terms this result of 25·07 as a "really glorious consummation." "For," says he, "a length of 25·025 British inches, or practically the sacred Hebrew cubit, is exactly one ten-millionth of the earth's semi-axis of rotation." We grant that it is nearly so under Professor Smyth's erroneous summation; and if we allow the polar axis of the earth to be, as Professor Smyth wishes it to be considered, 500,500,000 British inches. But in Captain Clarke's latest and very careful observations on the subject, the polar axis is not this measurement. Mr. Taylor considered the pyramidal inch and cubit to be an exact segment not of the axis of the earth at the pole, but of its axis at 30 deg. lat., where the Pyramid stands, and where the earth's diameter is, of course, much greater. Yet Mr. Taylor and Professor Smyth made both axes, though very different in length, fit exactly to their numbers. Mr. Taylor made out further the "sacred cubit" to be 24·90, whilst he found the pyramidal cubit 25 inches, each different, but both supposed to be revealed from on high. To account for this, Mr. Taylor states, and despite all geological or other authority, that "the smaller (24·90) is the sacred cubit" which measured the diameter of the earth before the flood; the one by which Noah measured the ark, as tradition says, and the one in accordance with which all the interior works of the Great Pyramid were constructed. The larger (25) is "the sacred cubit" of the present earth, according to the standard of the Great Pyramid when it was completed. Such contradictions and inconsistencies, said Sir James, entirely upset the whole wild and extraordinary theory of the base of the Pyramid being a standard of lineal measurement, or a segment of any particular axis of the earth. Mr. Taylor and Mr. Smyth seem to think that if, by multiplying any object in length, they can run out the calculations with a long tail of 0's, it proves something exact and certain. The lecturer stated and showed that if Sir Isaac Newton's standard of the "sacred cubit" be true, and if the polar axis of the earth be, as fondly and theoretically imagined by Professor Smyth, 500,000,000 pyramidal inches, then the brim of the lecturer's hat being exactly one-half of Sir Isaac's 'sacred cubit,' that brim was in measurement one 20,000,000th of the earth's polar axis. The lecturer pointed out other common objects by which such lines of 0's could be similarly obtained, if that availed aught. Sir James next discussed some others amongst the many matters which Professor Smyth fancifully averred to be typified in the Great Pyramid. The so-called Queen's Chamber, has a roof composed of two large blocks of stone leaning against each other, making a kind of double roof. This double roof, and the four walls of the chamber, according to Professor Smyth, count six, meaning the six days of the week, whilst the floor counts, as it were, a seventh side to the room, "nobler and more glorious than the rest," and typifying something of a "nobler and more glorious order"—namely, the Sabbath. It is surely difficult to fancy anything more unfounded than this strange imagination. Yet Professor Smyth holds that he is thus able to point out that the Sabbath had been heard of before Moses, and that, in the above typical way, he finds unexpected and confirmatory light of this—a fact, which he holds, is of "extraordinary importance, and possesses a ramifying influence through many departments of religious life and progress." He holds also that the corner-stone, so frequently alluded to by the Psalmist and the Apostles, is a symbol of the Messiah—is the head or corner-stone of the Great Pyramid, which, though long ago removed, may yet possibly, he thinks, be discovered in the Cave of Machpelah, though why or wherefore it should have found its way to that distant and special locality is not suggested. Professor Smyth, continued Sir James, holds the Great Pyramid to be in its emblems and intentions and work "superhuman," and "built under the direction of chosen men divinely inspired from on high for this purpose." If divine, the work should be absolutely perfect; but, as owned by Professor Smyth, it is not entirely correct in its orientation, in its squareness, etc.—all of them matters proving that it is human and not divine. It was, Professor Smyth further alleges, intended to be a standard of measures for all times down to these latter days, "to herald in some of those accompaniments of the promised millennial peace and goodwill to all men." If the Pyramid, continued the lecturer, is a divine miracle of this kind, then the Arabian Caliph El Mamoon upset the supposed miracle a thousand years ago when he broke into the King's Chamber and unveiled its contents; for the builders intended to conceal its secrets for the benefits of these latter times, and for this purpose had left a mathematical sign of a cross-joint on the gallery, by which some man in the distant future, visiting the interior, should, according to Professor Smyth, detect the entrance to the chambers, and which secret sign Professor Smyth himself was, as he believes, the first to discover. In relation to the Great Pyramid, as to other things, we may be sure that God does not teach by the medium of miracle anything that the unaided intellect of man can find out. The lecturer concluded by a series of remarks, in which he pointed out that at the time at which the Great Pyramid was built, probably 4000 years ago, mining, architecture, astronomy, etc., were so advanced in various parts of the East as to present no obstacle in the way of the erection of such a noble Royal Mausoleum as the Great Pyramid undoubtedly is.

'The CHAIRMAN (Professor Kelland) said he was afraid that the meeting could not enter into a discussion that evening on the paper just read. Professor Piazzi Smyth was undoubtedly entitled to a reply, and others might have something to say on the subject. He would, therefore, suggest that they should hear any remarks which the Professor might wish to make at present, and that the discussion should be resumed at another meeting.

'Professor PIAZZI SMYTH, in reply, said—The essay which we have heard this evening is both on a very important subject, and by one of the most celebrated and versatile men of the present day. If the result of this essay should be the elucidation of more truth than has hitherto been known in the world, no one will be better pleased than myself. But, if that result should not follow, and if the essay should be found unable to withstand the necessary test of subsequent careful examination, the responsibility of having uttered its errors will attach to the author alone; and there it may be left. So far then, Sir, touching the advertised subject of the paper, according to its title as duly printed in the billet for the business of this evening, and to which I would gladly confine myself. Yet I cannot pretend to be altogether unaware that—such printed title of the paper notwithstanding—my humble name has occupied, unfortunately, a large share of the lecturer's attention. The primal reason for this would seem to be, that in the course of a special investigation a short time ago, I ventured to discuss the merits of two little paragraphs in a certain book which was printed and published to all the world, in the beginning of last year. Those two paragraphs crossed the path of my investigation in so positive a manner that it became absolutely necessary for me either to disprove them or yield to their domination. In all this there was nothing unusual, for the same sort of thing takes place, more or less, in every scientific inquiry; and, in fact, in that same investigation of mine there was a host of other books submitted to a precisely similar line of criticism. Yet what was said, in one short section, touching only the two paragraphs already mentioned, has excited an amount of comment (from their author) which is perfectly extraordinary. The first of those remarkable paragraphs of his, contained the assertion that there are "pyramids," yea, even "magnificent, though rude" pyramids in Ireland; and the second stated that, "from the very "earliest historic periods in the architecture of Egypt, "Assyria, Greece, etc., down to our own days, circles, "single or double, and spirals, have formed, under various "modifications, perhaps the most common fundamental "types of lapidary decoration." Now, my argument with regard to the first paragraph was, that whereas the shape of a pyramid is defined truly and mathematically to be a solid, whose base is a regular rectilinear figure, and whose sides are plain triangles, with their several vertices meeting together in a point, that name of *pyramid* could by no means be even approximately applied to Newgrange, the chief of the so-called Irish pyramids, because, when restored from its present to its original condition, it is shown to have had a circular base, and a top, not pointed, but very broad and flat. And my argument touching the second paragraph was that, in so far as referred to the Great Pyramid of Jeezeh—the example, *par excellence*, of the very earliest historic period in the architecture of Egypt—(with the other and later countries I had nothing to do)—in so far, I say, as referred to the Great Pyramid of Jeezeh, circles, single or double, and spirals, did not by any means form the most common fundamental types of its lapidary decoration. There has, indeed, been a suggestion thrown out that I was altogether mistaken; and because—as gathered out of my own book, *Life and Work at the Great Pyramid*,—on a certain absurdly small feature of the whole building, usually known as the "granite leaf," there is, on one part of it, a peculiar, little, something or other; a something, whether for ornament, or mechanical use, or symbolical purport, and when put on no one can tell. In order that the presently known facts of this most minute and peculiar little something or other may be easily apprehended, I have placed on the table a model of the whole "granite leaf," one-fifth of the natural size; and I shall be much obliged to any gentleman who will show how there can be got out of the small, smooth, slightly-raised, semi-ovoid patch you see thereon, enough engraved figures of "circles, single or double, and spirals, to form perhaps the "most common fundamental types of the lapidary decoration" of all the twenty-one acres of the Great Pyramid's well-dressed stone surface. Further than this, I cannot say that anything of importance has been brought forward trenching on the real grounds of my views concerning the unparalleled and very exalted character of the design of the Great Pyramid. Much attempt has no doubt been made to break down distinctions, and confound it with other pyramids, and even with remotely pyramidal structures; and much attempt also to undervalue it itself, merely because some of its (the Great Pyramid's) describers have erred, or had imperfect data of observation and measurement to go upon. But this is an old, old story. It is what the Pagan Egyptians began; what the Greeks went on with; then the Romans; then the Mohammedans; and lastly, even some modern Europeans—some few men, who are in a sad state of darkness as to modern science; the final and highest principle of which is, not to presume to give out perfect and infallible dicta, but rather to know well about the limits of error attending every numerical datum. But the Great Pyramid has outlived all these unwise attacks; and it still stands, amongst many other of its features, the highest of all buildings, and with a most important meaning in that particular height. Several cathedral spires of Europe have been intended by their architects to be higher; but, from foundations failing, or other reasons, they have invariably stopped far short of their aim; leaving, therefore, the earliest of existing buildings, the Great Pyramid, still in all its noble superiority as well as scientific significance; and making it more than ever a wonder and admiration to all the intelligent of mankind in every land. After standing, indeed, thus for 4000 years, the Great Pyramid is now shining forth with a brighter character and purer fame than ever it did before. And only think, Sir, what that implies. To withstand the trials of 4000 years of persecution, and oppression, and criticism, and judgment, and yet to come out fairer and more precious after all! I can only hope that the same may be said, after the lapse of even a tenth part of that interval of time, for the characters of some who, in persistently endeavouring to undervalue that primeval structure, know not what they do. (Applause.)

'Owing to the time occupied by the above subject, the reading of the other papers on the programme was postponed till next meeting of the Society.'

MEETING OF APRIL 20, 1868.

As reported in the SCOTSMAN *of April* 21.

'Last night, at the tenth and last ordinary meeting of the Royal Society for the season—Professor Playfair in the chair—a paper was read by Professor C. Piazzi Smyth commenting on two papers on the Great Pyramid lately printed in the "Proceedings of the Royal Society," one of which was the production of a native of Sweden, and the other was written by Sir James Simpson. Both of these papers were alleged by Professor Smyth to be based upon needlessly imperfect data. The first was written in entire ignorance of the progress of pyramidal research during the last three years; and he had placed himself in correspondence with the author, who had confessed his ignorance of all recent Pyramid exploration so honestly that he could do nothing else than present him with a copy of his recent work. The other essay (Sir James Simpson's) extended to no less than twenty-six pages, of which twenty-four pages were filled with hardly anything else than the severest animadversions on the Great Pyramid and himself—(laughter)—his name being printed no less than eighty-five times—(laughter)—and there being no fewer than seventy cases of extracts in inverted commas from his writings, besides many quotations similarly derived, but without commas. After maintaining that an opposite meaning was ingeniously given in the essay to the majority of the extracts from his writings, than that which he intended them to convey, Professor Smyth proceeded to contend that Professor Simpson was mistaken in his ideas regarding the construction of the Pyramid, and said that he saw no prospect of that gentleman ever coming to understand the true construction of that noblest work of the world 4000 years ago, and how vastly it differed from all those tumuli which he insisted on comparing it to and confounding it with, unless he would himself ascend and descend the ancient Pyramid, climb up the steep steps of its colossal masonry, and measure each step carefully from top to bottom. He went on to defend the accuracy of all his measurements, and to insist that the authority upon which Sir James Simpson relied for the statement that an embalmed human corpse had been found in the coffer was not reliable.

'Sir JAMES SIMPSON, in reply, said that Professor Smyth was mistaken in supposing that he attacked the Pyramid when he objected to his reasonings upon the measurements of the Pyramid. It reminded him of Bishop Warburton always maintaining that those who attacked his reasonings in his work on "The Legation of Moses" were not condemning him but Moses. He was at perfect liberty to express his opinion on the Pyramid question, and he had not knowingly mis-stated the meaning of any quotation which he had made. Sir James then noticed a few of the measurements of Professor Smyth, maintaining that the calculations which he had founded upon them were inaccurate, and that it was an erroneous method of procedure to take the mean of different measurements. Such a method of procedure, he alleged, was childish ; it was a species of mathematical aberration, and it ran through the whole of Professor Smyth's book. Professor Smyth, he believed, was one of those men who would compute millions, but could not calculate the smaller figures.—(Laughter.)—He had not heard a single sound man say otherwise than that the whole of Professor Smyth's theory about the Pyramid was a series of the strangest hallucinations, which many weak women believed, and a few womanly men, but no more. He had talked about it to a great many arithmeticians, and he had found that they scoffed at it and despised it.

'Professor SMYTH, in reply, said that, with regard to the two papers in the "Proceedings" of the Society to which he had confined himself, he was already in correspondence with the author of one of them, and hoped to arrive at some approximation of opinion with him ; but with regard to the author of the other paper, he thought that it was proper that he should not attempt to hold any discussion with him, until the Council should have had abundance of time to inquire into the matters that he had laid before the Society, and had arrived at some conclusion or report upon them.'

REMARKS ON THE LECTURE OF JANUARY 20,

As it was reported in the DAILY REVIEW *of January* 22 ; *in a Private Letter from* W. P., *of date January* 23, 1868 : *not written for publication.*

'Had it been my part to reply to Sir J. Y. Simpson's discourse,—I should have felt it necessary to commence by expressing a difficulty in knowing how far I was expected to take his observations as mere pleasantry, jocular play, not intended to be answered. For, of this character, much of it certainly seemed.

'His mode of representing the coffer's condition;—his complaint of the errors of modern measurers as a charge against the Great Pyramid;—his complaint of (what is just the *token of design* in) the coffer, as intended for a *capacity*-standard, that it is unamenable to linear measurement; and his complaint that if linear measurement were intended, the intended standard would be repeated many times, all over the coffer and Pyramid; whereas it is considered by scientific men to be important, *not* to repeat a standard more often than is necessary,—in order to avoid either excessive labour in making copies identical, or else confusion from their casual differences. Though the force of this reasoning could hardly be admitted by one who gives out, that *no* errors of workmanship or measurement ought ever to be made, even by ordinary mortals, if they wish to be considered as doing a thing at all respectably.

'Again, his taking the mere *admitted* limits of error in modern imperfect measures under adverse circumstances of dilapidation, hurry, insufficient means, etc.,—and declaring that all persons are *agreed* that these errors are not that, but *original variations* in the structure itself;—surely all this is mere joking, and strange waste of time for the Royal Society of Edinburgh.

'The same gentleman's other mode of attack too; the picking out the *mistakes* (whether arithmetic, or inductive, or of scientific data), *mistakes* made here and there by those who have elaborated the whole theory; but only here and there amid many striking things where there is no mistake; yet mistakes which, in some cases, when rectified show the shrewdly-guessed idea in a still stronger light. He calls up such mistakes as if the leading outlines of the theory had no other supports! As if such mistaken proofs were not *generally* committed, by most advocates of a very new theory in an unusual subject, amid all the more or less sound evidences of it!

'If so charged, *I* could feel no confusion or difficulty; it is merely what has been abundantly seen in all scientific progress. Special cases of mistaken arguments or data, raised in defence of, or constructing some theories,—which nevertheless do not perish with the failure of these weak proofs, but on the contrary survive and are firmly held, on quite other and independent grounds. Though I would rather take my stand on the simplest of all platforms, and ask, "Where is the man who never made a mistake?"

'The lecturer's argument about Khaliph Al Mamoon, and that his power to injure the Great Pyramid is proof that that building was not of Divine origin,—is entirely false; but it is a fallacy which has wide application. If valid, it would prove that the Mosaic ark also was not built by Divine command (supernaturally given in real human words proceeding from no mouth of born man, and supernaturally shown to Moses' eye in the Mount),—because that ark was captured by the Philistines.

'Also that the Temple of Solomon (with all its contained vessels), was not built by supernatural Divine command,—because it and they were all either destroyed or carried away to Babylon, and used there to drink out of, in honour of idols! But that libation was an expensive one to the

Khaliph who performed it. That act brought the handwriting on the wall, and cost Belshazzar his kingdom and his life, with the thenceforth wasting away of the city.

'Or again, that Christ could not be Divine, because nailed to the cross. There were those who asserted that as a reason at the time, to wit, the Chief Priests. God answered the taunt not then, but afterward by resurrection and ascension.

'Or indeed the same principle, similarly applied, proves the non-existence of God; for he allows man to deny his existence, and to say and do what they please against Him (but mind! only during the day of grace)—without any hindrance. The argument indeed proves nothing; and it is *so manifestly faulty* (like most of the lecture) that it is really not worth these few pages of hasty writing.

'The hat-measuring performed by Sir J. Y. Simpson before the assembled Society, is beneath contempt. Any one after he *knows* from others, what the 10^{-7} part of the earth's radius is,—can find just as many ridiculous or ignoble portions of matter in art or nature as he pleases, and which shall be, as nearly as he may rather coarsely wish, a round fraction of that same cubit.

'If he wishes to illustrate his argument by a *really* parallel case, let him take any other structure as large and equally simple in its parts as the Great Pyramid, and then let him show the existence of as many and as close coincidences to cosmical data. But *even then*, he would not have done half the work; for, there must be also a coincidence of a system of symbols already recognised as existing among men in the part of the world where the structure existed.

'My own views of the Great Pyramid's Divine origin (views very gradually formed, and cautiously, during the last two years), are such that I could afford to make a present to the opposite party of the whole of the metrical facts! and rest the case entirely on the Scripturally-fixed and recognised sense of the symbols of form, etc., which compose the structure;—combining that species of evidence with the evidence of the direct allusions to it in Scripture, and showing that these two distinct classes of evidence *exactly* coalesce; *without* the (certainly powerful) additional evidence of *measure*, *equally* coalescing with the other two!

'I so much feel the importance of this subject, that if I were to be laid aside by illness or by the approach of death, I should (in a natural sense) deeply regret that these papers were not completed;—not for any ingenuity in them, nor that *I in the least* expect that the subject itself will bring earthly honour to any one handling it—(it appeals to too small a minority, and always must do so;—not one in a hundred is likely to receive its truths, or even to understand them properly);—but because I think I can just *point*, to a deal of truth which some true Christians may then study and appreciate; but which at present they seem not *to notice*, and chiefly for want of a pointing finger. I had no idea that the subject would expand so much, under such study of it as I have been able to give. W. P.'

REFLECTIONS ON THEORIZING;

AND ON THE SUPERNATURAL;

IN A PRIVATE LETTER FROM W. PETRIE, Esq.

The most plausible of Sir J. Y. Simpson's objections, on January 20, is, that the theory he condemns is not held without variations, by each individual who advocates it, and that even the same advocate has occasionally modified his theory. What an accusation! that an important theory, having many parts and ramifications, and while its data remain very incomplete, is not seen quite correctly in all its parts as soon as its leading principles can be grasped. What a suggestion! that the necessity of modifying the details of a newly developed theory makes its principles stand self-condemned. Where would modern science be if its advocates had acted on these principles? The great engine of modern science, *induction*, itself pre-supposes and necessitates the building up of a true theory, out of a mass of what are, at best, mere probabilities in its favour, continually confirmed by other equally slight probabilities, or set aside thereby. This process implies, as the almost necessary history of a theory ultimately adopted as beyond doubt, that it shall have been constructed out of what were originally mere probabilities (and very slight ones, nay, often only the best alternative among actual improbabilities), and that many of these may have been dismissed as erroneous in the process of maturing the theory.

What if the sacred cubit, in the course of such a process of theorizing, has been reckoned, by successively stronger probabilities, to be at first a few tenths of an inch, and then a few hundredths, different from the amount which further investigation makes to be the more probable truth? In any purely mundane investigation, no reasonable *savant* would think of taking such a circumstance as evidence of the fallaciousness of the general course of the development of the theory. Indeed none more fully recognise the justness of this, and habitually take extreme advantage of the liberty of frequently modifying a theory, than that school or party (the rationalistic) which is most forward to consider the exercise of this liberty to be a proof of the untenableness of the whole theory, in such a case as that of the Great Pyramid,—where the theory happens to be adverse to those of the school referred to.

The alternative theories which Sir J. Y. Simpson and other opponents offer, leave far more unreconciled features in the structure and in its known history,—more that is unaccounted for, and improbable, than does the theory he opposes.

That theory generally, is addressed and offered to those only who already admit that a Divine command supernaturally given, is a *vera causa ;* as science terms a cause recognised already to have existed in the case of other known phenomena, and, therefore, in a certain sense at hand or available for the present hypothesis. Those who openly deny this, are quite consistent in pronouncing this whole Pyramid theory to be fanciful. To such this theory is not offered: a previous question must be settled with them, namely, the *existence* of supernatural words and acts from the Creator to His intelligent creatures: or,—according to a symbol chosen by Him as the most applicable—the existence of the Presiding Potter and the fact of the touch of His finger, as a *vera causa* in the history of the Cosmos, over and above the mere relation of the wheel (of nature) on which His work is being performed as on a continuous and uninterrupted basis.

It is no discredit to the Cosmos nor to its Creator,[1] that He designed and intended it to produce results by connexion with,—and not as isolated from,—His direct action. It is no discredit to the potter's rotary apparatus, that it should be intended for his own direct use; not to work as a clock in his absence, but to produce its results by the combination of the intelligent and varied touch of his hand with the monotonous rotation of the apparatus itself; while, moreover, that too, is continuously sustained by his foot—see Postscript. This idea is skilfully embodied in the characteristic structural features which pervade the Great Pyramid.

I would rather err by sometimes ascribing too much to the finger of the potter, than by refusing to admit his

[1] Though this has been asserted as if self-evident,—*Essays and Reviews*, and similar works.

action, except by the foot, at the treddle; or by altogether ignoring his presence at the wheel.

WILLIAM PETRIE.

P.S.—A HASTY NOTE, OR OUTLINE OF THAT ON WHICH A PAPER IS IN PROGRESS.

ROTATION—scripturally symbolizes Nature,—τροχὸς τῆς γενέσεως, Natural law, the fixed order of creation, or cosmos itself; therefore, in contrast to

RADIATION = Supernatural power,—ἀπαύγασμα τῆς δόξης (Heb. i.), the effulgence of the Shekinah,—Almighty power that originated nature itself, and which keeps it going;—and which does not desert it, but continues pervading it,—*using Nature* in *combination* with this direct supernatural power, this 'finger of God,'—the ray of his own outstretched arm and hand,—as when the Spirit brooded on the waters of the *cosmos* to produce organic *life*. The principle is symbolized in nature itself by solar radiation working in concert or unison with terrestrial rotation.

This all-important principle, that the supernatural and the natural, in *combination*, produce a *result*, and that the result is not produced by either separately, is very plainly illustrated in Scripture, by the symbol, several times employed,—the *Potter's Wheel*. His *wheel*, with rotary action (natural law, Eccles. i. 3 to 13), works in unison with the action of his finger or hand (the supernatural, termed in Scripture 'the finger of God),' which is at right angles, any way, to the rotary motion, —an expressive symbol of their absolute non-correlation; these, combined, produce the vase. The presiding potter, meanwhile, himself sustains the motion of his wheel, *on* which also he does his work of production. 'Thou art the potter, we are the clay,' nature, *cosmos*, is the wheel, and thus the potter's hand represents supernatural power working in the midst of natural law:—as in composing the first organic germs of the various vegetable and animal generations, and the first human couple, and as also seen subsequently, in angelic ministrations, and the working of the Holy Spirit in Divine inspiration, etc.

The rationalists wish to say that the vessels are produced by the potter's wheel *without the potter*, and that the idea that his wheel cannot do *that* without the hand, is a libel on it, and is as much as to say that it does not fulfil its purpose!

They call the application of the potter's hand during the rotation of his wheel an *interference* with { the laws of nature. / the wheel.

W. P.

LETTER No. 3.

FROM THE DAILY REVIEW EDINBURGH NEWSPAPER,

For May 18, 1868.

OPINIONS OF A CIVIL ENGINEER.

'SIR,—In the report of the last meeting of the Royal Society of Edinburgh, which appeared in your impression of the 21st ult., I find that Sir J. Y. Simpson is represented to have said, regarding the work in three volumes lately published by Professor C. Piazzi Smyth on the Great Pyramid, "that throughout the whole of it there ran a series of the strangest hallucinations, which only a few weak women, and perhaps a few womanly men, would believe; and that he had spoken to a great many engineers, mathematicians, and others, and he found them scoffing at and despising it."

'Now, Sir, excepting by reputation, I have not the pleasure of knowing Sir J. Y. Simpson; and I certainly cannot boast of the wide experience which may have fallen to his lot in respect to acquaintance with "weak women" and "womanly men," whatever that expression may mean; nor could I possibly predicate what such persons might or might not be generally inclined to believe; but if, guided by their instincts for what is true, they have attained to faith in Professor Smyth's general deductions, I say it is so much the better for them, and that it is certainly no argument against the correctness of his conclusions that they may have done so. The insinuation intended to be conveyed by the sneer is unworthy either of the reputation or social position of the utterer of it, and must be received as a proof that he finds himself much in the position of the celebrated defendant who, having no case, endeavoured to make up for that awkward deficiency by abusing the plaintiff's attorney.

'But further, Sir J. Y. Simpson appears to have had the misfortune to fall into the company of a number of engineers, mathematicians, and others, whom he describes as much given to "scoffing" and "despising," and I fear, further, that in his eagerness to reciprocate their opinions he has had the additional misfortune also to have acquired their peculiar method of communicating them. Now, Sir, I must say that it strikes me that the description he gives of the singular mode in which these friends of his have conveyed their opinions to him respecting Professor Smyth's book is such as to lead me very much to doubt the value of their conclusions. "Scoffers" and "despisers" are not, as a rule, people of sound judgment; but what can possibly be said in defence of the judgment of any one so ill judged as to quote their judgments in aid of any views of his own!

'I happen, Sir, to be a member of the institution of civil engineers. I am also a member of the institution of mechanical engineers, and I have had, as such, considerable experience as a producer in connexion with one of our staple manufactures, and amongst gentlemen who, with no ordinary combination of enterprise, practical ability, and scientific attainment, conduct some of the great industrial concerns of the nation, and I have the pleasure to inform Sir J. Y. Simpson that, whenever I have had the opportunity amongst my fellow-engineers, mechanicians, or manufacturers, of explaining the views enunciated by Professor Smyth I have always found respectful attention paid to them, and in many cases a singularly appreciative adhesion to them, and especially to the system of weights and measures proposed by Professor Smyth, and founded on the metrical standard and system of the Great Pyramid. And even those who are not at present prepared absolutely to adopt the Pyramid standard, greatly appreciate the extreme beauty and simplicity of the Pyramid system of mensuration which the Professor has so ably elucidated.

'For myself, I can only say that, having given the operative conclusions of Professor Smyth my most careful consideration, I believe them to be correct: and I hope to live to see the day when both the standard and system of measuration he has deduced from the Great Pyramid will take the place of our present unworthy standard and cumbrous system; and as far as in me lies, I for one will do all in my power to promote what I should consider such a happy practical consummation of the labours he has bestowed upon this great subject,—I am, etc.,

'HENRY J. MARTEN.

'WOLVERHAMPTON, *May* 14, 1868.'

APPENDIX 3.

M. WACKERBARTH, OF UPSALA, ON THE GREAT PYRAMID;

AND

THOUGHTS THEREON BY MR. W. PETRIE.

The former extracted from the Proceedings of the Royal Society of Edinburgh, Session 1867-68.

'*On the* GREAT PYRAMID OF GIZEH, *and* PROFESSOR C. P. SMYTH'S *Views concerning it. By* A. D. WACKERBARTH, A.M., *of Upsala. Communicated by* PROFESSOR KELLAND.

'The author gives a detailed statement of the theories of Professor Smyth, as given in the Transactions of this Society, vol. xxiii. part 3. He then, after heartily commending the zeal and diligence of the Professor, brings forward objections to some of his views. 1. As to the metron or unit of linear measure. Mr. Wackerbarth objects that this measure was utterly unknown to the ancient Egyptians—appearing in no Egyptian document or monument whatever, nor in any ancient writer who describes the condition of the Egyptians.[1] Mr. Wackerbarth collects from the

[1] Both M. Wackerbarth and the writer of the abstract seem here to state, as a new discovery of the former gentleman's, with regard to *one* of the scientifically deduced standards of the Great Pyramid, a quality or feature which is the already recognised and pervading characteristic of the *whole* of those standards; and forms an important step in the chain of evidence leading towards a recognition of the august authorship of the original design of their containing and component Monument. See chapters 22 to 32 of this book.—C. P. S.

materials most accessible the following table of weights and measures in use amongst the ancient Egyptians (corrected for the recent measure of the cubit of Karnak, made by Sir Henry James).

Long Measure.									English Inches.
1. ⲧⲏⲃ (Digitus),									0·78925
2	1. Κόνδυλος,								1·4785
4	2	1. ϣⲟⲡ (Palmus),							2·957
5	$2\frac{1}{2}$	$1\frac{1}{4}$	1. ϭⲡⲁⲧϩ (Vola),						3·69625
6	3	$1\frac{1}{2}$	$1\frac{1}{5}$	1. Kubdeh,					4·4355
8	4	2	$1\frac{3}{5}$	$1\frac{1}{3}$	1. Διχάς,				5·914
11	$5\frac{1}{2}$	$2\frac{3}{4}$	$2\frac{1}{5}$	$1\frac{5}{6}$	$1\frac{3}{8}$	1. Fitr,			8·13175
13	$6\frac{1}{2}$	$3\frac{1}{4}$	$2\frac{3}{5}$	$2\frac{1}{6}$	$1\frac{5}{8}$	$1\frac{2}{11}$	1. ⲉⲣⲧⲱ (Σπιθαμή),		9·61925
28	14	7	$5\frac{3}{5}$	$4\frac{2}{3}$	$3\frac{1}{2}$	$2\frac{6}{11}$	$2\frac{2}{13}$	1. ⲙⲁϩ I cubit),	20·699

Itinerary Measure.

1. ϩⲟⲟⲩⲧⲡ (Στάδιον),			606·875 ft. Eng., or 623 ft. Swedish.
60	1. ⲡⲟϩ ⲛ̀ⲡⲣⲱϣ (Σχοῖνος),		6·9 English miles, or 1 Norwegian mile = $\frac{1}{10}$ degree of equator.
150	$2\frac{1}{2}$	1. ⲁⲟⲩⲣ . . .	17·2 English miles, or $2\frac{1}{2}$ Norweg. miles = $\frac{1}{4}$ degree.

Land Measure.

1. Ἄρουρα = (100 cubits)^2 = 10,000 square cubits = 0·683 English acres = 0·557 Sw. Tunnaland.

Wine Measure.

1. Ἀρυστήρ = about $\frac{1}{2}$ pint English, or about $\frac{1}{10}$ Swedish kanna.

Weight.

1. ⲉⲙⲛⲁ (Mina) = 8304 grains = 1·1863 lb. av., or 1·278 Swedish Skålpund.

'2. The author then, finding no such measure as the metron, goes on to say :—

'"It is really very unpleasant to reject a beautiful and ingeniously carried out theory which one wishes to believe, but I cannot, however, arrive at any other conclusion than that Professor Smyth's very interesting account of an ancient Egyptian standard of length, exactly equal to a ten-millionth of the earth's axis of rotation, and the entire system of weights and measures that he has deduced from this imaginary unit, requiring accurate knowledge of the earth's dimensions, figure, and density, several centuries before the age of Abraham, and the supposed connexion of the English measures with these, are things purely mythical. On the other hand, that the Pyramid and its contents were really the standards of the ancient Egyptian kingdom, is or seems to me an opinion not destitute of probability; but I cannot think that those standards were constructed with any reference to the dimensions of the earth, but that they were arbitrarily chosen quantities, intended to represent on a tolerably large scale the size of those portions of the human body which their names indicate : *e.g.* ⲙⲁϩⲓ (*cubitus*), the arm from the elbow to the point of the middle finger; ϣⲟⲡ (*palmus*), the breadth of the hand, etc. etc.

'"In fine, Professor Smyth's method, namely, that of multiplying or dividing by quantities, for the introduction of which it would be hard to give any satisfactory reason, —as, for example, the number 366, with which the base was divided, and which is neither the exact number of days in a sidereal year, nor anything else, and again, the very uncertain number, 5·672 (the earth's density),—appears to me in the highest degree dangerous. How deceitful results thus obtained can be, I shall now, in conclusion of this discussion, attempt to illustrate by a few examples; and before doing this, I beg to state that, entertaining the highest respect for Professor Smyth, I have not the least intention of making light of his work, though this method of reasoning, being analogous to the *reductio ad absurdum*, will unavoidably at times have something of that appearance.

'"*Ex.* 1. It is a historical fact, that the present English weights and measures are not of any very great antiquity. At the time of the Norman Conquest the yard was about 39·6 of the present English inches, a little longer than the French metre; and the foot, accordingly, 13·2 modern inches, a little greater than the Paris foot. In the year 1101, King Henry I. determined the yard by the length of his own arm, and that is the determination which the present yard is intended to represent; and that yard, moreover, has never been, even in modern times, defined by any fraction of any of the earth's dimensions, but by its proportion to the length of the seconds' pendulum at $51\frac{1}{2}°$ latitude, and of this yard the English foot is the third part. Now, a degree of the equator is just 365,260·524 feet. Divide a thousandth part of that number by the length of the sidereal year in solar days, 365·256358, and we have 1·0000114; that is to say, if we take a thousandth of a degree of the equator, and divide it by the number of days in a sidereal year, we have an English foot as nearly as a powerful microscope can determine it. *And yet it is certain that this is purely accidental.*

'"*Ex.* 2. If I take 10,000 times *e*, the base of the hyperbolic logarithms, and multiply it into the quantity, which in the lunar theory is called *g*,—that is, the ratio of the difference between the moon's and its ascending node's mean motions to the moon's mean motion,—and divide the earth's polar radius by the product, the result is the length of the Pyramid's side. But are we to suppose that the Egyptians forty centuries ago were acquainted with the lunar theory, the earth's compression, and the use of logarithms, and, moreover, took this clumsy method of perpetuating their knowledge? Is it not far more probable that the architect simply determined that the length of the side should be 444 cubits, and accordingly measured out that distance?

'"*Ex.* 3. Again, if instead of taking Colonel Vyse's measure of the base, we accept for that quantity the mean of half a dozen of the most trustworthy travellers' statements,—for instance, Vyse's, the French Academicians', Caviglia's, Wilkinson's, Lane's, and Davison's—we have for the length of the side in English feet a number which expresses in millimétres the mean height of the barometer at Upsala. But will any one maintain that the dimensions of the Pyramid were intended by its builders to have any reference whatever to that interesting constant?

'"*Ex.* 4. If we multiply together one-tenth of the side of the Pyramid's base, the length of the line joining the middle of the side and the apex (that is, the height of each one of the four isosceles triangles that compose the Pyramid), and the modulus for the common logarithms, the result is 3420, the constant of lunar parallax, for which, in Burg's tables, the value 3420″·96 is given.

'"*Ex.* 5. Lastly, if the side of the base (763·81) be divided by the hyperbolic logarithm of π (the circumference to diameter), and that quotient again by the ratio of the force of gravity at London to the force of gravity at the Pyramid (lat. = 30°), (1·00188), the result is 666!'"

ON PROBABILITIES OF COINCIDENCE;

OR,

THOUGHTS SUGGESTED BY M. A. D. WACKERBARTH'S STRICTURES ON THE RECENT METRIC THEORY OF THE GREAT PYRAMID:

BY WILLIAM PETRIE, ESQ.

31*st March* 1868.

Mr. Wackerbarth's table of ancient Egyptian measures is in perfect accord with all that Professor Piazzi Smyth says. The agreement of both parties on the most ancient Gentile cubit (20·7 British inches) is certainly striking, seeing the very different sources and dates from which each derived the fact: Wackerbarth from a wooden measure discovered at Karnak, Professor Piazzi Smyth chiefly from the interior of the Great Pyramid, inaccessible to the makers of the Karnak measure, and separated from it by 700 years. Yet the agreement is to the twenty-thousandth part of the whole quantity: this is a difference of only a small fraction of the probable error of each party's source of information.

So far, then, there is a striking confirmation of Professor Smyth's views instead of any stricture on them; the divergence relates purely to *the other* standard of measure, the metron of 50·05 British inches; or, what is really the same in point of theory, a standard of 25·025 British inches, the precise half of that length: and this is the form in which it ought more justly to be discussed, as it pervades both of Piazzi Smyth's publications, while the metron or two-fold form of the standard is only partially assumed in the earlier and smaller book, *Our Inheritance*, and is not rested on at all in the larger and more recent publication, entitled *Life and Work at the Great Pyramid*, the half length appearing, on further investigation, to have been more probably the original—or at least one of the original forms of it. Mr. Wackerbarth's objection[1] to the theory on which this is deduced, is, that it rests on coincidences which may, too possibly, be merely casual or fortuitous. To illustrate this, he cites examples of curious numerical coincidences that are indisputably nothing more than what are commonly called casual. The mention of these as a caution merely, would be laudable and useful;

[1] This objection having been before the minds of many, it is desirable to examine carefully whatever seems the most ably argued form in which it has been put. On this principle of selection, Mr. Wackerbarth's paper takes precedence of one by Sir J. Y. Simpson, contributed to the same learned Society (R.S.E.), as this last-named paper abounds in mere invective, and in erroneous assertion, rather than in solid argument.—W. P.

but let us consider how far the application of such examples of coincidence to Professor Piazzi Smyth's theory of a 25·025 inch standard is deserved or sound.

His first example is, that a 360,000th of earth's equatorial circumference is 365·2605 British feet, being the number of solar days in a year, with an arithmetic difference of only one 100,000th part, which is within the probable error of our knowledge of the datum involved. This is certainly a rare coincidence: if a computer were to devote the working hours of each working day to search physical science as quickly as he could for such really unconnected coincidence, the average rate at which he would discover them would be just about one in each year so spent. We therefore feel much indebted to Mr. Wackerbarth for it, though it be nothing but a curiosity. But, it confessedly *means* nothing: hence its weakness as bearing adversely on the Pyramid theory. Had it been a precise *decimal* fraction instead of having 36 as the submultiple of earth's equator, there would have been more approach to meaning in it; and had it been a circle of earth's orbit around the sun, it would have had a distinct meaning, and besides, would give a round number of feet in each day's march of the earth; but to take the attributes of one natural circle and to attach them to a different natural circle, is a process adverse to any theory that could assume the coincidence to have been intentional. The coincidences of the Pyramid theory differ in these respects from such mere coincidence as the above: for particulars see the reply to Sir H. James.

Mr. Wackerbarth's observations (see Proceedings of the R.S.Ed. 1867-8, and our page 450), on the arbitrary and modern length of the British yard, prove nothing as to *the inch*, which is spoken of as an ancient measure in Professor Smyth's work; nor even as to the foot of 12 such inches, of which our non-venerable yard (as he truly describes it) is made, for convenience, to contain an integral number.

Mr. Wackerbarth's other examples of remarkably close coincidences between unconnected numerical data have the same defect, of being dissimilar from the leading Pyramid coincidences,—both in the manner already explained, and in being artificially complex, and of heterogeneous composition; hence they have no real force as against the Pyramid theory in question. Only one part of that theory (that involving the mean density of the earth), is somewhat behind the rest in extreme simplicity, and in such ostensible meaning and reasons—seen for instance in the tokens of earth's radius and sun's mean distance,—as to be in some small degree open to an imputation of improbability of intention, as in the case of the above-named mere coincidences; and, to this extent, that part of the theory does stand less firmly than the rest. Its author and advocates quite recognise this. They have never said, nor do they mean, that every point of the theory is equally secure from mistake, and that it admits of no improvement or modification: only that there is a strong probability (equal to what is in practical life termed a certainty) that the leading ideas of it are, to some important extent, real;—worthy of careful improvement, rather than, as Sir James Simpson seems to think, worthy of hostile efforts; as though similar coincidences, when undesigned, were found a hundred times more commonly than they are, and as though the suggestion of a communication of superhuman knowledge were so Bibliolatrous as to deserve *no quarter*. Yet that key is the really *existent* key in human history;—a *vera causa* (according to Baconian phrase), and being so, it is as much entitled to be fairly tried as any other key—in this and in *any* physical problem,—and much more so entitled, than some keys which *savants* are fond of trying, unabashed (millions of ages, for instance), which are not *veræ causæ* of effects similar to those (organic, geologic, or otherwise) that they wish to explain thereby.

But further,—*why* do simple and intelligible coincidences constitute stronger proofs than such as those cited by Mr. Wackerbarth, viz., complex ones, and such as do not carry with them a manifest connexion of ideas? Chiefly because the *number* of facts, amongst which a single close coincidence is to be searched for, is vastly greater in the latter case than the former. For example, Mr. Wackerbarth's cases of fortuitous coincidences with Pyramid quantities are so vaguely and capriciously connected with the subject, and are so complex, that they can be manufactured by myriads, and the abstract probability of the *intention* of coincidence (shown by the closeness of agreement, to say about $\frac{1}{3000}$th part, as in the instances given), must be divided by about a myriad. This makes $\frac{8}{7} \times 3{,}000 \div 10{,}000$ to 1,[1] or $\frac{1}{3}$, being less than even chance, if we would express the true value of the coincidence as representing a probability of design or intention in the matter. Try our Pyramid theory by this test. We have a coincidence say to $\frac{1}{3000}$th part. But before proceeding, let it be distinctly understood that this does not mean that much of certain error or want of coincidence, as Sir J. Y. Simpson seems to misconceive it, but that the difference is not proved to exist at all, except from necessary imperfections of work "made by hands," and of its measurement by limited powers of sight, and of the needed accuracy of knowledge of the physical data required. These fractional deficiencies of perfect coincidence being then thus understood, we have, in the Pyramid theory, a coincidence say to $\frac{1}{3000}$th part on an average of the cases in question. This coincidence is from a class of facts limited by connexion with the subject and with fellow-coincidences; so much so, that the choice does not lie among fifty such facts; nay, not among a dozen such—perhaps

[1] The mathematical reason for inserting the constant $\frac{8}{7}$ would lead to a digression which the reader will not practically need, as it affects both sides equally.—W. P.

half a dozen, when we take into account their agreement with traditions, etymologies, and symbols in the structure; as was more fully explained in a paper read by Professor Piazzi Smyth before the Royal Society of Edinburgh on 16th March. We have therefore to divide $\frac{8}{7} \times 3000$ by about 6, making about 571 to 1, call it 600 to 1 (because the data of the calculation can only be very vaguely assumed), as representing the probability (so far as the degree of coincidence shows it) that the cause of this coincidence was design or intention, as distinguished from a purely fortuitous fact. Moreover, if there be two or three such coincidences arising out of *the same* theory, their probabilities must be separately estimated on the above principle, and then multiplied together; two such of 1000 to 1 each making a combined probability of a million (1,000,000) to 1.

The above is a rough illustration of the mode of mathematically or quantitatively expressing the correct mental impression of probability, as to intention and causation having originated any such remarkable coincidences as those above referred to, in dimensions, time, qualities, etc. Such an impression is inherent in the common sense of mankind, when it is matured by experience of the practical facts of the cosmos, and is not warped by prejudice or interest. The impression referred to (of the immense power of evidence in two or three close coincidences when they occur in the circumstances above stated), has so good a foundation in truth, that such strictures as we are examining, suggested by *apparently* similar examples, without design or causation, useful as they are in their right places, are yet out of place when allowed to apply to coincidences such as those of the Pyramid theory, the conditions of which are shown to be highly different in degree, though similar to a superficial view.

Some may ask if the divisor in our above-named fraction representing the probability, should not be reduced to the number of instances individually tested for a coincidence by the seeker of the coincidences, instead of being, as we may seem to have taken it, the number of instances which the seeker *might* test if he had unlimited time? To this we reply that we have taken a number between the two, as being the truest; because the mind does, in reality, test a much wider range of the worse or more obviously non-coincident class of instances (by a rapid intuition or practised expertness unconsciously exercised in rejecting them), than those instances merely which it is conscious of testing, and which are chiefly the small percentage of better sort of coincidences. This fact is well illustrated by the action of the eye, which, in a second of time, can unconsciously notice literally ten thousand objects before it; such as seeds of many colours, shades, shapes, and sizes, evenly shaken out upon the sufficiently small meshes of a sieve, and can single out from amongst them the one, or the few, that seem likely to reward further examination by yielding some desired coincidence of size or position on the meshes, etc. But, in justification of the *general* truth of the preceding comparisons of Mr. Wackerbarth's coincidences with those of our Pyramid theory, let it be remembered by those who think the assumed denominators to be unduly large, that this error, if it exist, is very much compensated by its affecting both sides; and thus the comparison would not be essentially altered by limiting the field of instances searched in both cases to those which were somewhat consciously noticed by the mind, in its search for coincidences.

But the calculations, to find out the probability for or against the existence of any causative connexion, should be completed by the introduction of another element as a co-efficient; namely, the *intrinsic improbability of such a causative connexion existing*, irrespectively of the question of more or less coincidence of any sort, and reckoned thus purely from *general considerations of the nature* of the hypothetical cause, or rather of all the alternatives available as hypothetical causes, whose respective probabilities are intended to be noticed. This element cannot well be reduced to a numerical expression, for it depends on a wide range of the relations of things. Still the estimate must be made. No error in the estimate of this element can be so great as that which would be caused by the entire omission of the element from the calculation, on the plea that it is difficult to form an estimate of it, and that any estimate will be undesirably vague. The common sense of most men unconsciously makes the estimate, even while they refuse to do it intellectually or numerically, saying it cannot be done.

Let us now take one of the examples of coincidence given by Mr. Wackerbarth, as inviting a comparison of probability (or rather of improbability) with that of our Pyramid theory, as regards the possible existence of a causative or intentional connexion between the coinciding facts. We have here to balance such an improbability as the particular event of the *present* modern mean atmospheric pressure, estimated in a *particular* measure, and by the height of a *particular* liquid—mercury—at one out of thousands of somewhat celebrated *spots on the earth*, ancient and modern—Upsala—and other things, as being known to and connected with the builders of the Pyramid, and selected by them for commemoration by the Great Pyramid. We have, I repeat, to balance such an improbability as that, against another improbability: let us review, then, its leading features also. That the design of the Great Pyramid—involving the leading geodetic and solar phenomena; expressive not only of what has taxed the best powers of modern science to ascertain, these most noble physical data themselves, but by their aid as well as by a device of structural angle, expressing with redoubled emphasis the idea of rotation in union with radiation;[1] and the structure being the greatest and earliest of remaining monuments, and in its features and history showing strong opposition and contrast to its surrounding idolatry—had the same origin as our Sacred Scriptures and the stone cairns and various other structures therein recorded with precise measurements. Seeing that these Sacred Scriptures and these structures therein mentioned are similarly opposed to surrounding idolatry; are of similar date; and declare similarly, the sun to be set for (symbolic) 'signs, and for seasons, and for days and years,' and 'the breadth of the earth' to be a noteworthy object by the Designer and Declarer of the words of those Scriptures. Some persons would say that there is *primâ facie* much probability rather than improbability in the last-named connexion; but we do not ask our opponents to go so far; we only ask that it be granted that the connexion is *incomparably* less improbable than any causational connexion with the Upsala fact above named.[2] We would suggest that the probability is about 1,000,000,000 to 1 against the latter, and (not to favour our own theory too much), 10 to 1 against the former. Then the probability ratios ($\frac{1}{3}$ to 1, and 600 to 1) already above obtained in each case respectively, and derived from the testimony of coincidence alone, should be reduced by dividing by these figures respectively. The ultimate probability thus obtained, after duly including *both* the nature of the coincidences and the abstract improbabilities of their being causatively connected, stands as follows:—

$$\frac{1}{3} \times \frac{1}{1,000,000,000} \text{ to } 1, \text{ or say } \frac{1}{300,000,000} \text{ to } 1$$

[1] A consideration of this and other great scriptural symbols in the structure is in preparation for publication.—W. P.

[2] It will be seen that the superhuman origin of the Scriptures is here taken for granted; so far at least, as concerns the scriptural facts above named. We assume this because we only address ourselves, on this occasion, to those who admit it. For others, the scriptural question must be previously settled, as an entirely separate subject.—W. P.

against such a coincidence as Mr. Wackerbarth mentions, being causative or intentional,

$$\text{and } 600 \times \frac{1}{10} \text{ to 1 or 60 to 1}$$

in favour of even one isolated case of coincidence, such as our Pyramid theory presents, being causative or intentional in the original design of the structure.

There are some other elements,—such as, the small fraction expressing the mean or probable error of workmanship originally unavoidable; and, more discrimination between the one least improbable kind of connexion, in the case under consideration, that of design and work of a living intelligence, and any other kinds of causative connexion between coincident quantities, much more improbable, as the alternative to its being supposed a purely fortuitous coincidence. Such additional elements are properly necessary to the theoretic completion of any such computation of probabilities as the preceding, but these are not noticed above, because in the existing state of our dimensional knowledge of the Great Pyramid they would not materially affect even the separate probabilities of intention in the two classes of coincidence here reviewed; certainly not the comparative result here established.

We shall be reminded that Mr. Wackerbarth distinctly professed *not* to give cases similar in degree to that of our theory, but professedly extreme cases, by way of *reductio ad absurdum.* We have not overlooked this, and our object has been to show that the same sound principles which show the absurdity of those cases, show the contrary of ours: and that where there is a question of probability (as facts of human knowledge ever must be, however certain they may be practically called) the demand to consider them in the same light as less probable cases, constitutes a species of *reductio* in which the *absurdum* consists solely in the process of *reductio.*

W. PETRIE.

APPENDIX 4.

EXTRACTS

FROM AN

ACCOUNT OF MAESHOWE IN ORKNEY,

BY JOHN STUART, SEC. SOC. ANT. SCOT.

CONTAINED IN VOL. V. PART 2, OF PROCEEDINGS OF SOCIETY OF ANTIQUARIES OF SCOTLAND (1864).

'IT was thus plain that Maeshowe was a great cham-'bered barrow, and that it had been constructed by first 'building on the surface of the ground the chambers and 'gallery, then covering the structure with a layer of small 'stones, and finally heaping over this a layer of earth from 'the adjacent surface, and from the (circular) trench by 'which the barrow is surrounded.'—P. 251.

'Chambered tombs have been found in many countries, 'or a series of chambers, in the centre of mounds of earth 'or of cairns of stones, approached by long passages. In 'Denmark, structures of this sort are called giants' graves; 'they are found throughout the north of Europe; they 'occur frequently in Brittany; and very remarkable ex-'amples have been found in some of the Channel Islands. 'But all the chambers now referred to, are formed by the 'imposition of covering flags on walls also formed of flags, 'as in the simple cromlechs, without any appearance of the '*horizontal arch* which is a distinguishing feature of the 'central chamber at Maeshowe.

'The horizontal arch formed by the projection of one 'stone beyond the other, till the conveying walls are 'spanned by a single flag at the top, is found in the under- 'ground chambers in every part of the north of Scotland, 'as well as in the chambered cairns and "brochs." It is 'also the distinguishing feature of early Irish remains of a 'similar character, the cloghauns, the underground cham- 'bers in raths and other places, and in the chambered 'cairns at Newgrange and Dowth on the banks of the 'Boyne.'—P. 251.

Maeshowe in Orkney and Newgrange in Ireland compared.

'I paid a visit to Newgrange shortly after my eyes had 'been familarized with the arrangement of Maeshowe, and 'the likeness between the two structures was so remark- 'able, as to leave no room for doubting that the same idea 'had suggested both.

'In each there was a long narrow gallery, issuing in a 'central domed chamber, from which three lateral crypts 'branched off, although the structure of Newgrange is 'much ruder and less developed than the other. This, 'however, may be partly accounted for by the differences 'of material at Maeshowe; the builders got abundance of 'slabs in the neighbourhood, split up in such regular forms, 'that the work has the appearance of ashlar, without being 'touched by a tool; while at Newgrange the builders had 'principally to deal with large shapeless pillars and small 'boulders. Here the gallery leading from the outside to 'the central chamber, and which measures 63 feet in 'length, is formed of large upright stones, covered with 'flags of immense size. The central chamber is formed by 'a series of upright pillars partially sunk into the ground, 'on which the converging walls of the dome are supported, 'and the whole covered with enormous quantities of small 'boulders exactly on the principle of the chambers in Picts' 'houses. The top of the dome is 19 feet 6 inches from the 'floor. From the entrance to the inside wall of the chamber 'opposite, measures 18 feet, and between the extremities 'of the crypts on the right and left, the distance is 22 feet. 'The chamber itself is of an irregular shape, not above '8 feet in diameter in any direction.'—(See *Archæologia*, vol. ii. p. 254.)

'The chamber at Maeshowe was about 15 feet square at 'bottom, and the top of the dome, when complete, was 'probably about 20 feet from the floor. If we include the 'cells, the structure measured about 28 feet across, and 'nearly 20 feet in the opposite direction. In each of the 'recesses at Newgrange is an oval stone, slightly hollowed, 'which may have been used for the same purpose as the 'raised slabs on the floors of the cells at Maeshowe. It 'seems probable that in both cases the remains of the dead 'in some urn or cist were placed here.

'The cairn at Newgrange was protected by a circle of 'great standing stones, which surrounds it at some distance 'from its external limits, as the ditch does at Maeshowe; 'and if the cairn of stones were removed, it would leave a 'structure, consisting of a narrow alley of standing stones, 'leading to an irregular circle, with three offshoots of 'similar pillars, which may suggest some points of analogy 'between it and groups of pillars not under cairns, such as 'Callernish.'

Late and Early Visitations.

'When Newgrange was examined in the seventeenth 'century, it was found to be much in the same state as it 'now is. We are told that underfoot there was nothing 'but loose stones of every size in confusion, and among 'them a great many bones of beasts and some pieces of 'deer's horns; and with the exception of a quantity of 'bones and teeth of the horse, and a small fragment of 'a human skull of unusual thickness, which were found 'in the *débris* in the chambers, no relics were found at 'Maeshowe. There is reason to believe, however, that 'in both cases there had been early visitors of these 'mounds, who had left the chambers in confusion, after 'having rifled them of everything of value which they 'contained.

'The rifling of tombs was carried on to a great extent in 'the middle ages; and the temptation to it was great, from 'the practice which long prevailed of burying, with the 'departed, jewels and ornaments of great value.

'That the Danes ransacked Newgrange and the other 'tombs of the Kings of Tara in the year (861 A.D.), we are 'assured on the authority of the Irish annalists (*Annals of* '*Four Masters*, by O'Donovan, pp. 497-498), and we shall 'see that Maeshowe suffered a like visitation (between 800 'and 1100 A.D.). Whether a gold coin of Valentinian '(370 A.D.), and another of Theodosius (390 A.D.), found 'on the outside of Newgrange, or the two ancient gold 'torques, a golden chain, and two rings, found in digging 'a little to the west of the entrance, ever formed part of 'the treasure which it contained, we cannot say; any more 'than whether the silver ornaments found near the bay 'of Skaill, and now in our Museum, formed part of that 'great treasure said to have been carried off from Maes-'howe,' (and containing coins, Saxon and Cufic, of dates 'from 887 A.D. to 945 A.D.).

Pictish, Pelasgic, and other comparisons.

'The idea which suggested the arrangement of the 'chambered tomb was only a modification of that which 'governed the builders of the "Picts' houses." One large 'central chamber, surrounded by a group of smaller ones, 'and approached by a long narrow passage, is the description 'tion applicable to this latter class of buildings, exemplified 'by those at Quenterness and Papa Westray in Orkney, 'and Kettleburn in Caithness.

'In all these the passage to the internal chambers is so 'confined as to imply a crawling posture in one requiring 'access, and the communication between the chambers is 'equally confined. We find the same narrowness of pas'sages in subterranean chambers in Ireland. . . . They are 'the characteristics also of the African wigwam of the 'present day. . . .

'The idea of the sepulchral chambered cairn was much 'the same, only that the internal cells seem to have been 'fewer in number, while not unfrequently there was only 'one large domed apartment in the centre, approached by 'a narrow passage; as in the striking group of cairns at 'Clava on Nairnside, described in our "Proceedings" (vol. 'iii. p. 47); and as in the chambered cairns at Yarrows in 'Caithness, opened by Mr. Rhind, and described by him 'in the *Ulster Journal of Archæology*, vol. ii. p. 100.

'In this partial similarity of arrangement between the 'sepulchral and the living chambered cairns, we recognise 'a feature in which they agree with one class of the early 'Etruscan tombs, which have their internal arrangements 'constructed in imitation of a dwelling chamber, with fur'niture, like the apartment itself, cut out of the rock '(Fergusson's *Handbook of Architecture*, p. 259). Many of 'these Etruscan tombs consist of five or more chambers, 'vaulted on the horizontal principle, in the centre of 'mounds of earth, which are surrounded by a circle of 'standing stones.

'The curious specimen of early Pelasgic art, called the 'Treasury or Tomb of Atreus, one of the kings of Mycenæ, 'bears a strong resemblance to the plan of some of our cham- 'bered cairns. In it is a great circular chamber, in a mound 'of earth, nearly 50 feet in diameter, vaulted on the prin- 'ciple of horizontal layers of stones projecting the one beyond 'and above the other, till one small stone closed the whole, 'and made the vault complete. This chamber, approached 'by a long gallery, has on one side of it a small crypt cut in 'the rock, which appears to be the true sepulchre. . . . The 'central chamber at Maeshowe was unpaved, and although 'Mr. Farrer, at my suggestion, made some digging into the 'soil, no deposit was found, nor was there any appearance 'of the soil having been disturbed. So that here also, the 'lateral crypts may have been the real sepulchre.

'From what has been said, it seems likely that Maeshowe 'must be regarded as the work of a race who long pre- 'ceded the coming in of the Norse population. . . .

'To how much earlier a period than the ninth century A.D. 'we are to assign the erection of Maeshowe, it would be 'difficult to say. We may learn, however, from Adamnan, 'in his Life of Columba, what was the practice of burying a 'Pict of some importance soon after the middle of the sixth 'century A.D. (viz., in the place where he died he was buried 'by his companions, they raising a *cairn* of stones upon 'him).

'This was no doubt an extemporized funeral, but we 'know that the cairn was one of the recognised monuments 'of great men, both in regal cemeteries and on battle-fields, 'in Pagan times, in Celtic Ireland, till the establishment of 'the Christian religion, when all such modes of burial were 'abolished (450 A.D.)

'Among the Celtic people of Ireland certain spots were 'set apart as regal cemeteries. Of these one is on the 'banks of the Boyne, and contains the cairn of New-'grange, and many other monuments of the kings of 'Tara. It would seem that Stennis (in sight of Maeshowe) 'had been a spot of similar consecration at some remote 'time for the people of Orkney, or, more generally, of the 'north of Scotland, although it is not easy to suggest a 'reason for its selection.'

APPENDIX 5.

WEIGHTS OF THE GREAT PYRAMID AND OF THE EARTH;

APPROXIMATELY.

FOR THE WEIGHT OF THE GREAT PYRAMID.

Linear elements.		*In Pyramid cubits.*
Vertical height of Great Pyramid, ...	=	232·52
Inclined height of Pyramid face, ...	=	295·72
Side of square-base, ...	=	365·24
Transverse thickness of general casing-stone film,	=	4·

CUBICAL RESULTS APPROXIMATELY.

Cubic Pyramid cubits in whole Pyramid, .			=	10,339,850
Hollow space of Grand Gallery, .	=	2,625		
Hollow spaces of regular character elsewhere, as in chambers and passages, . . .	=	2,625		
Subtract, ...		5,250		
				10,334,600
Subtract x for irregular occasional hollows, and add same for extra specific gravity of lining of King's Chamber, ...				0
				10,334,600
Subtract cubic contents of casing-stone film of a specific gravity peculiar to itself, . . .			=	861,952
Remaining cubic contents in Great Pyramid, all of one supposed nummulitic limestone density, (though x portion of it is Mokattam stone),			=	9,472,648

Specific gravity of casing-stone, or Mokattam stone, (See *Life and Work*, vol. iii. p. 177.)	=	0·367
Specific gravity of nummulitic limestone, . . (See this book, p. 234.)	=	0·412
		Pyramid tons.
861,952 cubic cubits of 0·367 specific gravity, . (See *Life and Work*, vol. iii. p. 175.)	=	395,420
and 9,472,648 cubic cubits of 0·412 specific gravity,	=	4,878,414
Sum, or concluded whole weight of Great Pyramid, approximately, ...	=	5,273,834

Weight of Earth approximately.

Polar Diameter, . . .	=	20,000,000 Pyr. cubits.
Equal Diameter, . . .	=	20,070,000 ,,
Arithmetical mean of the two radii,	=	20,035,000 ,,
Cubic Pyramid cubits in Earth's bulk viewed as a sphere of above mean radius, .	=	4,210,791,200,000,000,000,000
Pyramid tons in the same, or $\frac{1}{4}$ more, being necessarily of the Earth's mean density or mean specific gravity, . .	=	5,263,489,000,000,000,000,000

Weight of the Earth more accurately; in a private letter from W. P. to C. P. S.

Equatorial radii, . . .	251,116,200 251,039,700	British inches. ,,
Arithmetical mean, and practically the true mean, . . .	251,077,950	,,
Polar radius, ...	250,250,000	,,

(Ratio, say 300 : 301 accurate within limits of error.)

You have worked with what, I presume, are the above data (as given in *Life and Work*, as the best ultimate determination of modern science), only expressed in sacred cubits.

Mean Equatorial radius, ... 250,827,120 s. c. inches.
Polar radius, ... 250,000,000 s. c. inches.

That is to say,

Mean Equatorial *diameter*, in sacred *cubits*, . . 20,066,170
Polar *diameter*, in sacred *cubits*, ... 20,000,000

In using these data you have (doubtless for brevity, and as appearing likely to be practically exact enough for your purpose) taken the *mere arithmetical* mean of the two diameter lengths, and reckoned that a perfect sphere (by the regular formula $\frac{\pi}{6}\cdot D^3$) with this diameter would be sufficiently equal, practically, to an oblate spheroid of those different equatorial and polar diameters. But in truth this does a gratuitous injustice to your latest Great Pyramidologic thesis $\triangle$ mass $\times 10^{5\times 3} = \oplus$ mass (or weight).

The precise mode of reckoning in this case would be (as you are well aware, but I repeat it for the sake of distinctness, and to save you the trouble of turning to the formula, and to enable you to check *me* if I make an oversight therein),—to take, for the *true mean diameter for the purpose* of giving the mass, not $D+\frac{1}{2}a$, as you took, but exactly $D+\frac{2}{3}a$, D being the polar diameter, and a the *excess* of the general equatorial diameter beyond that of the Polar.

We get it, as I note below.

	Expressed by the radius R.	Expressed by the diameter D.
A sphere is, ...	$\frac{4}{3}\pi\cdot R^3$ or $\frac{4}{3}\pi\cdot R\cdot R\cdot R$	$\frac{\pi}{6}\cdot D^3$ or $\frac{\pi}{6}\cdot D\cdot D\cdot D$
And, if the sphere be, as the earth is, increased by a small amount equatorially, or in two of its three axes, with a true ellipse for any trans-polar section (an ellipse being a figure with every one of its ordinates the same multiple of that ordinate to its inscribed circle), then the effect increased weight of the earth's oblateness will		

	Expressed by the radius R.	Expressed by the diameter D.
be exactly in proportion to its circumscribed cube similarly made oblate, that is, increased in its two equatorial axes.		
That is to say (with the notation used on my previous page), ...	d = the difference of the radii	a = the difference of the diameters.
Neglecting the a^2 fractions because they are smaller than can be practically estimated,—just as we neglect the second differential of a quantity,—we have thus—		
As a sphere is, ...	$\frac{4}{3}\pi R \times R \times R$	$\frac{\pi}{6} \times D \times D \times D$
Then a spheroid is, . .	$\frac{4}{3}\pi \cdot R \times (R+d) \times (R+d)$	$\frac{\pi}{6} \times D \times (D+a) \times (D+a)$
or, . .	$\frac{4}{3}\pi \times (R+\frac{2}{3}d)^3$	$\frac{\pi}{6} \times (D+\frac{2}{3}a)^3$
In other words, *the mean radius, for the purpose* of calculating *cubic bulk*, .	is $R+\frac{2}{3}d$	$D+\frac{2}{3}a$.

Applying this correction to your own calculation of earth bulk, it makes it to be

5,270|976,000|000,000|000,000 } Pyramid tons, instead of what you
5,263|489,000|000,000|000,000 } give without the correction.

But as your diameter used was not the exact arithmetical mean, there is still a difference to be corrected.

No longer referring then to your working,—I give here, under the results of my own perfectly independent calculation, made many months ago, for the purpose of trying *very accurately* whether the space included in the whole King's Chamber, or in its lower course, or in one of the upper ones, had any relation to the bulk or weight of the whole earth; I did not carry on the examination very completely (I regarded bulk rather than weight of water and of mean earth),—and I found nothing worth noticing to you.

Taking earth $= \frac{4}{3}\pi \times R+(R+a)^2$, being the true formula for the bulk of an ellipsoid,—I brought out earth (IF an exact ellipsoid) by the radii given on first page—

= 66,	086,000	000,000	000,000	000,000	cubic inches, British measure.
= 65,	888,000	000,000	000,000	000,000	cubic inches, Pyramid measure.
=	4,216	830,000	000,000	000,000	cubic (a. c.) (Pyramid) measure.
=	5,217	,037,500	000,000	000,000	Pyramid tons weight. Sp. gr. 5·7.

But this gives an earth if it were pared down to the mean sea-level, and a perfect ellipsoid. We have then to apply some corrections. The atmosphere may be omitted as being a long way within the limits of our probable errors of observation; for it is only = 6 feet depth of earth's mean density, or $\frac{1}{1200000}$ of whole weight. Next,—taking the *water* in the world, *as it is*, simply, that is, occupying its existing space, light though it be; and making an allowance only for the projections of land above the mean sea-level: or, in other words, taking the sea-level at what it *would* be, if all the land were shovelled into the sea, so that no land could be seen above water; the effect of this would be to raise the mean sea-level only $\frac{1}{12}$ of a mile; if, as has been *roughly* estimated, the mean height of land is now $\frac{1}{4}$ mile above mean sea-level, and covers $\frac{1}{3}$ of earth's surface.

This would introduce, it is true, only a small correction into our calculations; $\frac{1}{12}$ mile in about 4000 miles radius, or $\frac{1}{48000}$ of the whole *lineally*, that is $\frac{1}{16000}$ of the whole, additional, in *cubic bulk*, and therefore, to see its effect on the weight, we must take the s. g. of this extra mass = 2·5 water and that of mean of all the rest 5·7 water, and say $\frac{25}{57} \times \frac{1}{16000} = \frac{1}{36480}$, or say = ·000028 more *weight* than is at present reckoned; and this correction will introduce a proportionate alteration in the calcalated mean specific gravity of the whole, in *some* (not in all) processes of calculating it.

Another correction needed, is, for the *supposed* more or less square form of any section of the earth in any plane of longitude,—that is, passing through each of its poles. But as this feature of earth's form is not well proved, it must be estimated at a *small* amount, to show the truth according to present probabilities. I suggest to add, therefore, only $\frac{1}{24}$ of the equatorial protuberance, or $\frac{1}{5000}$ of the earth's whole bulk on this account; that is $\frac{1}{7500}$, or ·000133 of earth's weight.

Then, adding these corrections, ·000028
+·000133
= ·000161, or say ·00016 of the whole = 843,000 | 000,000 | 000,000 Pyramid tons. Then, adding this to the exact sea-level spheroid at specific gravity 5·7, which

is	5,271,	037,500	000,000	000,000
		843,000	000,000	000,000
There results, as the final determination of the mass practically existing, as our earth,	5,271	880,500	000,000	000,000

Expressed in Pyramid tons.

Or say, omitting fractions which are certainly beyond probable error, 5,271 | 900,000 | 000,000 | 000,000 Pyramid tons.

On the Relation of the total Masses of Great Pyramid and Earth. $1:10^{5\times 3}$. By W. P.

This is one of those remarkable facts of which we have seen several examples in the constitution of the Great Pyramid; remarkable, inasmuch as that structure is found to represent cosmical quantities with much less differences than the supposed or probable errors in our data, by which we are enabled to try the question: while, nevertheless, this apparent precision has not been brought about by coaxing or intentional manipulating of the margins of uncertainty in the data used,—a process which others would naturally suspect to have been the cause of such undue precision in the results. But, be the inference what it may,—whether a large error be, in future days, discovered to vitiate the whole or not,—we are not responsible for consequences, but only for a faithful public statement of data, and of the facts in the constitution of the Great Pyramid, which the data reveal to us.

But the practical experience of a scientific mind will prevent so much surprise, or mistrust of physical inferences, theories, or laws, merely because of their coincidences being closer than the probable errors in the data used. The best and only true security against being misled by what is called mere coincidence, is to be found in noticing what degree of close or necessary connexion such result may have with a consistent and intelligible system indicated by other results: and this kind of proof exists firmly, in many instances, in our present unique and prolific theory of the Great Pyramid. Remarkable accuracy from uncertain data combined is known to occur, as a practical fact, much more frequently than would seem probable to abstract reason or popular expectation. I say this from individual experience, and setting aside the cases where the result depends on sagacity, or rather the natural instinct of a mind made for and devoted to physical investigation; an instinct as unaccountable to reason perhaps, as that by which a bird's bill, used with apparently no effort, nor even attention, weaves a nest much better than can be done by human reason with ten fingers and tools at command. This, in Sir Isaac Newton's case (as a signal example) is now known to have enabled him to propound quantitative facts, in several departments of science, to a degree of precision to which mere formal inductive reasoning would seem quite incapable of leading, on such rude data as were then available.

But to return to the Great Pyramid as a measure of earth's mass,—the working out of the results as they were hastily and provisionally made out and posted to me by Professor Piazzi Smyth, on his discovering the matter, carries its own evidence of impartiality, for, the coincidence (within $\frac{1}{500}$), though quite as near as could be expected from our data, even on the supposition, which probably ought to be granted,—that it was *intended* in the design of the Pyramid,—is yet not nearly so close an agreement as it would have been, had Professor Smyth honoured his thesis with a less rough process of calculating its application. His off-hand mode of calculating gave the following results :—

Earth's mass,	=5,263,000,000|000,000|000,000	in Pyramid tons.
Pyramid's mass,	=5,273,000	
Coincidence, to	10,000, being within $\frac{1}{500}$ of the whole quantity.	

But myself having, some months ago, required as a datum, the earth's mass according to the data acknowledged to be the latest and best, and the same as Professor Smyth had here used, but more deliberately worked out, with all minutiæ of correction,—I had by me the result, immediately available for the most impartial comparison. It is as follows :—

Earth's mass,	=5,271,900,000|000,000|000,000	in Pyramid tons.
Pyramid's mass,	=5,272,600	
Coincidence, exact to	700, being within $\frac{1}{7500}$ of the whole quantity.	

Details on all the above calculations accompany this sheet on a separate paper.

The Pyramid mass here adopted is taken as given by Professor Smyth's calculation of it, but corrected as seems needful for impartial probability's sake, by deducting 1200 tons (only $\frac{1}{4400}$ of the whole) for an undiscovered chamber, passage, and other vacuities. The reasons for the existence, position, character, and amount of which inferred spaces undiscovered, are given in a separate paper.

APPENDIX 6.

ON THE GREAT PYRAMID SCIENTIFIC THEORY, AND ITS CRITICS.

BY J. G.

Extracted from the EDINBURGH EVENING COURANT, *Saturday, May* 9, 1868, *and revised by its Author.*

'PROFESSOR PIAZZI SMYTH'S WORK ON THE GREAT PYRAMID[1] AND ITS CRITICS.

(*Communicated.*)

'The first volume of Professor Smyth's able work is in itself a valuable possession to a student. It contains the details of the everyday life of Professor and Mrs. Smyth (for his wife accompanied him to Egypt, as she had previously done to Teneriffe), from the time of their arrival in Cairo till their work in Egypt was finished; and it is written with an unflagging power of humorous, lively, and picturesque description. But, moreover, it is full of matter such as only an accomplished artist, a true scholar, and a pure and thoroughly practised man of science could have thrown into it. There is much of beauty in Dean Stanley's description of Egypt; but the pen of Professor Smyth presents such a picture as only a man full of science, and possessing the educated power of a painter, could draw.

[1] *Life and Work at the Great Pyramid;* During the Months of January, February, March, and April; in 3 vols. Edinburgh, 1867.

The river and its wealth, the fields, villages, and inhabitants, the monuments, grotesque or sublime, the heavens at morn or eventide, by noon of day, or noon of night, are all pictured to "the mind's eye" with a rare power, carrying with itself the feeling of truthfulness. The second volume, however, as it contains the data, in the form of measurements and their results in length, angle, and temperature, on which the theory set forth in the work rests, is in reality the most important part of the book, though to "the general reader" perhaps the least interesting. It is a remarkable product of a combination of patient research and unwearied assiduity in original observation. And let it be borne in mind that the observations and measurements are as little the work of a mere *dilettante* as are the calculations those of a mere scholar or antiquarian. They are the work of a man trained from his youth to the finest observations of this kind, and heretofore approved by his fellow-men of science, as an *expert* in such professional manipulations. He has been already tried by the only true and adequate jury as to his fitness for such work. A sentence pronounced again by that jury in a deliberate manner can alone be valid in shaking the credit of what Professor Smyth has set his name to as trustworthy. And we would boldly say that those who are training for work in the fields of scholarship, or history, or theology, would do well to learn from this volume how a true scholar, historian, and (in the real sense, like Newton) a catholic theologian, equips and girds himself for the work of founding his theory. The third volume contains the evolution and proof of this theory from the *data* set forth in the preceding volume. The theory is a consistent and beautiful *whole*, developing itself step by step from these *data*. It enables one to read in the proportions of the Pyramid, the scientific truth ascertained, or promising to be ere long ascertained, respecting the size and weight of the earth, respecting the time of the earth's revolution round the sun, and (as lately pointed out by Mr. Petrie) respecting its distance from the sun. It unfolds a stable and satisfactory system of metrology, based on the measureable length of the earth's polar axis, the mean density of the material substance of the earth, and the mean temperature of its surface; thus guaranteeing the superiority, in respect of appropriate earth-commensurability and trustworthiness, of certain traditional points in the long descended British system of metrology, and of the ancient Hebrew one, over the French metrical system of the revolutionary era. Moreover, besides indicating the period of the earth's revolution round the sun, and affording grounds for belief that there is indicated also symbolically the sacred division of time into sabbatical weeks, the Pyramid is seen to indicate on the sublime dial-plate of the "starry heavens" the precise date of its own construction; while here again its silent but demonstrative testimony is corroborated by the venerable voice of sacred history and of earth-wide tradition.

'No longer, indeed, does this noble geometric monument stand in all its primeval grandeur and perfection; but a deep meaning may be traced even in its gradual dilapidation. For thus alone could the alphabet of its language be learned, and the meaning stored up in it, spelled out step by step; thus alone could the real nature of its socket foundation be discovered, or the relation of its component layers of structural masonry to its symbology be ascertained; thus alone could the meaning of these things, and the kindred meaning of the august whole, be read into by the minds proved competent for the work.

'Of course, this theory has met with very various treatment. Some, acknowledging its power, have candidly confessed themselves placed in a dilemma by it; for, if true, it lands its recipient in the necessity of either admitting that the builders of the Pyramid possessed naturally a higher knowledge of physical science than is reached at this day, or that they worked on the strength of supernatural instruction and guidance. Others, again, without allusion to this dilemma, have generously acknowledged the value and beauty both of the measurements and of the theory. But still others, while laughing to scorn the bare idea of Divine inspiration and guidance in such a matter, have either argued against the facts on which the theory is based, in a very disallowable manner, or have made statements that are not warranted by the facts of the question. With a brief consideration of some of these things, and a short statement of our reasons for believing that the idea of a Divine interposition in this matter is neither irrational nor unwarranted, but rather highly credible, on the ground of analogy and an important final cause, would we close our remarks.

'An objection, according to a report that appeared in the *Courant* a short time since, has lately been brought against the system by which Professor Smyth reaches the ultimate result in his measurements, which, if valid, would utterly shake the theory by rendering its basis untrustworthy. This objection is directed against the system of taking the *mean* of several observations in order to reach the data on which the theory rests. This system is declared to be, in fact, worse than childish; to be a mathematical aberration, erroneous, equivocal, and delusive. Now, what we say is, that if such be the character of this system, Professor Smyth in employing it only errs in company with the great body of physicists, and that there is an end not only of his theory of the Pyramid, but to all the reliance heretofore placed on innumerable scientific results; nay, more, for in this case the very tables in the truth of which annuitants have been fain to trust are equivocal and delusive, and the fiars prices of every Scottish county are taken on a foolish and even dishonest principle. As for another statement, contained in the same report, that certain calculations of Professor Smyth were inaccurate, we have seen those special calculations gone over again and again carefully by another party, and found correct to their last figure; and we are informed that Professor Smyth has left the very sheet that was specially objected to hanging in the Royal Society Rooms, to be sifted and proved wrong by whomsoever may. Not unnaturally, the Professor asks, "Can Sir J. Y. Simpson "mean by such unsupported assertions to try to *stamp out* "the truth?" But, to close this part of the subject, we may state that we have beside us a list of the errata in his Great Pyramid work, printed by the Astronomer-Royal for Scotland, and this shows that, after all the criticisms directed against it, "the misprinting noted on pp. 2 and 3 of the list of errata merely of certain intermediate steps, for the sacred cubit, is the only error known to the author as discovered by Sir J. Y. Simpson and his friends." Again, a remark may be made on one or two out of many misstatements ventured on in two articles on *Life and Work at the Great Pyramid*, that appeared last summer in the *Saturday Review.* Thus the reviewer asked, "In the ruin-"ous state of almost every portion of the Pyramid, every "one must know how vague and indeterminate the best "observations must unavoidably be, as the huge discre-"pancies between the published lists suffice to indicate. "What are we then to think of measurements professing "accuracy to tenths of seconds, or two places of decimals?"

'Now we have here Professor Smyth's own answer to this in the following words:—"There are no observations "to tenths of seconds, and much less to hundredths of the "same, throughout all three volumes of *Life and Work.* And "further, I was at extraordinary pains in seeking out the "ancient unbroken surfaces of the Pyramid masonry to "observe upon in all important cases." The Saturday

Reviewer has about half a dozen similar instances of positive untruth, one of them being, that the great French national book on Egypt *does* show a *lid* to the coffer of the Great Pyramid. Now, a proper inspection of the book and plate referred to, shows that the lid meant by the reviewer belongs, according to the letterpress of the plate, not to the coffer of the Great Pyramid, but to a sarcophagus in a neighbouring tomb. Moreover, the plate representing the coffer of the Great Pyramid shows it, though under the name of sarcophagus, yet in fact and shape in the most unmistakable manner as a *lidless box* with all its angles sharply cut, and with not the faintest trace of any ledge or groove such as the reviewer would hint at. Such, then, is a sample of the inaccuracies occurring in a couple of brief articles, which would yet affect to explode and laugh out of existence a work which, we make bold to say, the reviewer himself could never have produced under the most favourable circumstances. He would need a different genius, as well as a different heart (and the two are closely connected), to produce such a work. And being thus led up to it, we would now venture a remark of our own on the bearing of the question—"Is the stone box of the Great Pyra-" mid a mere coffer or measure of capacity, as the great " French book unquestionably represents it?" Because,—on Professor Smyth's actual inspection of the coffer, it turned out to have a kind of ledge on which a lid might rest, but could not be immovably fixed,—therefore a conclusion has been come to by others and triumphed in, that the coffer is *only* a sarcophagus or coffin, and nothing more. Our remark is, that even if the proof of the box being, or having ever been, used for a sarcophagus were conclusively established (which is far from being the case, several things making it doubtful), still that would not invalidate in the very least Professor Smyth's metrological theory, based as that is on many well ascertained and measured numerical facts, with traces of consilient arrangements and commensurabilities. To our mind, it would rather give to that theory a deeper moral significance, and invest the entire subject with a deeper religious meaning. It would just add this to the theory, that the royal builder of the Pyramid, though deemed an atheist by the idolaters who toiled in his service, was enabled under the guidance of the great *Opifex Mundi* to rear for himself the grandest tomb that ever human being who hoped to live again was buried in. For, buried in that tomb and coffer, he lay among the symbols of creative and providential power, wisdom, justice, and bounty, continuing to preach of these attributes by his work so long as its symbolical meaning was remembered; even as he preaches by it again, in these latter days, after the knowledge of the symbols has been restored. But this we put merely as an alternative.

'In our opinion the idea of a Divine interposition in the planning and construction of the Great Pyramid, when closely contemplated as springing from all the facts and relations of the case, is perfectly rational and credible in the estimation of a rightly-instructed mind. Rightly-instructed mind, we say, for a man may be mighty in "midden" philosophy, and ignorant as a child in that great mother science of catholic and revealed theology, based on the grand design argument uttered by the Cosmos; on the wide testimony of universal history and tradition; and on that testimony of human nature to religion which is so inextinguishable, that it drives the very atheist Positivists into that ineffably sad idolatry of Humanity itself. It is on this grand testimony that the Astronomer-Royal for Scotland builds; and we rejoice to be of one mind with him. And well may Professor Smyth answer the scorners in the words of Newton to the scoffing Halley—"Mr. "Halley, I have studied the subject, and know the truth "of religion." And this not because we think the truth of religion, as the grandest historical element, is dependent on the truth of the theory as to the Great Pyramid, but because the principles involved in the full argumentation of this theory are among the principles of catholic theology according to our description of it; and, accordingly, whether the case of the Great Pyramid be one to which these principles are rightly applied or not, the principles themselves dare not be pooh-poohed. The self-called "advanced thinkers" of the archæological schools may scout them; but we hold, on the universal testimony of sacred and profane history, that man's story does not take its rise in a dunghill. Our creed in this matter is the same as Professor Smyth's—that blessed belief handed down in Scripture, and chanted by the grand choir of historians and poets. His theory, too, falls in completely with the grand strain. It points, on the ground of remarkable facts and coincidence,—to the Great Pyramid as an instance of those Divine interpositions which are known on the testimony of Scripture, corroborated by tradition, to have been made, as occasion called for them, during the infant ages of the world. Nay, more, for it points amongst others to one passage of Scripture (not, as the Saturday Reviewer says, because held to describe the Great Pyramid, but), because, since the passage evidently describes the earth symbolically, no other symbol except the Pyramid can be pointed to. The passage is—"Where wast thou "when I laid the foundation of the earth? declare if thou "knowest understanding. Who hath laid the measures "thereof, if thou knowest? or who hath stretched the line "upon it? Whereupon are the sockets thereof made to "sink, or who laid the corner-stone thereof, when the "morning stars sang together, and all the sons of God "shouted for joy?" But, moreover, the Great Pyramid, viewed in the light of the theory commenced by the late John Taylor, and further developed by Professor Smyth, is seen to be a peculiar one among other elements of prophecy, cast by Divine Providence as seed on the waters among the nations, to ripen in due time, and serve most beneficent ends in the appointed seasons. This is a theme treated germinally in such works as Bishop Horsley's "Preliminary Dissertation to his Sermons on the Resurrection of our Lord," and in the opening of De Quincey's "Essays on the Cæsars;" but it remains to be systematically and adequately handled. Some of its leading moments only can be here indicated. There existed, then, in the religious books of the ancient Persians, undoubted prophetic and apocalyptic elements, which certainly contributed along with other elements in the Magian system to form that character which fitted Cyrus and his Persians to punish the grossly idolatrous Babylonians, and free God's ancient people. Again, if ever there was a clear case of a Divine interposition of the more ordinary kind employed for great moral and religious ends, it may be seen in the moral and religious revival, such as it was, that took its rise in Pagan Greece in the person of the "daimon-possessed" Socrates, and all that sprang from the influence, example, and teaching, cast into society by that noble martyr. Still again, descending to Rome, it is matter of notoriety that the Romans treasured in the Capitol certain Sibylline books; that these, by very cogent arguments, can be shown *not* to have been favourable to polytheism, still less to pantheism; and that they not only fell in remarkably, in certain prophetical statements, with the Hebrew Scriptures, but influenced the conduct of leading Romans themselves, being supposed to apply to Julius Cæsar himself. All these things we devoutly believe to have been arranged and provided by God, even as we know from history that they formed powerful elements in forces that moved the cardinal events in human story. And is any one so blind as not to see that we live in times as momentous as any since those of the flood, excepting those years when the Lord of Glory himself dwelt on earth? For how many are ready to shout, Io Pæan! in the vain hope that at last the "vile superstition," as they call it, taught in Holy Scripture, and so marvellously supported, is doomed to a speedy extinction. Others are busily helping on this sure consummation, as they believe it, by advancing and fostering a strange philosophy, which (whatever lip-worship some of its sects may pay to revelation, yet in reality) takes man up at first as an ape-descended animal, reared in barbarism, and destined in the end (so far as this philosophy can show) only to make manure for the soil he sprang from. When forced to hearken to such degrading opinions, is it not a boon to be thankful for, when there is presented to our contemplation a most noble builded work, which proves how far removed from savageism its architects were, at a period when history and tradition alike testify, that man and the world had just emerged from an awful catastrophe? For in saying this we stand well supported, and defy any one to disprove, on the only valid and allowable ground—that of universal history and catholic theology—the reasonableness and credibility of God's interfering to instruct and guide an architect, who knew and worshipped Him, in the rearing of a grand symbolical building; suited, according to Divine foreknowledge, at least to stagger, and suggest wiser views to certain of the "advanced-thinkers" and rather too positive *à priori* philosophers of these latter days.

APPENDIX 7.

HEBREW SCRIPTURE CRITICISM,

IN CONNEXION WITH THE GREAT PYRAMID.

BY W. A. P.

ARTICLE 1.

(See page 537, vol. iii. of *Life and Work*.)

JOB iii. 13, 14.—'For now should I have lain still and 'been quiet, I should have slept: then had I been at rest, 'with kings and counsellors of the earth, which built *dry*, '*unfrequented places* for themselves.' חָרָבוֹת ('ghor-bāh';) (fem.) Literal Hebrew, precisely expressed.

Eng. Ver. has *desolate* places. The Hebrew root, חָרַב (ghōrah), he was dry, signifies dry, generally in the sense of mere absence of visible water; *e. g.*, 'the waters were *dried up*;' 'the Red Sea was *dried up*;' 'better is a *dry morsel* with contentment,' etc. But sometimes it signifies more strongly, a thing deprived of its invisible moisture, desiccated, parched; *e. g.*, 'I was like an owl of the *desert*' (fem.); 'in the day the *drought* consumed me' (mas.); 'my bones are burnt with *heat*' (mas.), Job xxx. 30; 'a shadow from *the heat*' (mas.)

Hence—from the absence of water implying cessation of vegetable growth, and therefore of animal dwelling—the the word is, even more frequently, used figuratively for *desolation* as regards vegetation or human dwellings than it is literally. Sometimes it is used even for killing (Jer.

l. 27); '*slay* all her bullocks;' *i.e.*, take away their (liquid) blood which is their life.

The word ('ghor-bāh') in Job iii. 14, is probably meant in *both* senses: a place where the mummified body is left to retain its dry and shrivelled state, and a place at the same time apart from human dwellings.

It is quite a canon, in Scripture criticism, that its author, the Holy Spirit, intends us to take *the whole* of the customary or conventional scriptural sense of a word, except so far as the context or other Scripture may limit that extended sense.

The description last above would apply equally to rock tombs or to a pyramid, and not to other burial places; but the term *built*, instead of *hewn* or *digged*, points strongly to the idea of a pyramid rather than to the only alternative, a rock tomb which was generally *digged* as a catacomb, with a well-like entrance, or at most, only hewn out of the side of a rock; not essentially *built* above it also, as is here stated.

Further, a pyramid is—by absence of light, and by length and narrowness of passages—less fitted for human dwellings than the rock-tombs such as surround them, and are continually used by the living; and therefore the idea '*desolate*' applies also, rather more forcibly to a pyramid than to a mere tomb.

ARTICLE 2.

(See page 537, vol. iii. of *Life and Work.*)

Lamentations iii. 16.—'He hath set me in *dark places* as they that be dead of old.' מַחְשָׁךְ (magh-shāhch) mas. The root חָשַׁךְ ('ghah-shach'), he was dark, signifies primarily *to restrain*, hold in, and *chiefly with reference to light*, because its being restrained restrains all human doings and organic life.

The same root is used in Lamentations iii. 2, 'He brought me into *darkness*;' iv. 8, 'Their visage *is blacker* than a coal;' v. 17, 'For these things our eyes are *dim*.' And Psalm cxliii. 3: 'The enemy hath persecuted my soul, he hath smitten down my life to the ground; he hath made me to dwell in *darkness* as those that have been long dead. Therefore is my spirit overwhelmed within me, my heart within me is desolate.'

As in Job iii. 14, so here, the only two places that can be meant, agreeably to Eastern custom, are a pyramid or a rock tomb; and, of the two, the sense of the word, either radical, to restrain or shut in, or derivative, to darken, shut light out,—is more completely applicable to a pyramid than to a rock tomb.

But there are also other expressions in the context that indicate a pyramid rather than a tomb. Blayney's translation is quoted in some places here. It has the merit of being quite impartial testimony, for he does not perceive the allusion to a pyramid. Parts of verses 4 and 5 refer to a mummified state equally in *either* place of burial. 'My flesh and my skin hath he made old; he hath encompassed my head so that it is weary' (Blayney, p. 394). But verse 5, 'He hath builded upon me' (Blayney)—against me (Eng. Ver.)

7. 'He hath hedged me round about that I cannot get out; he hath made my chain heavy.'

8. 'Yea, when I cry aloud he hath obstructed my prayer' (Blayney).

9. 'He hath blocked up my way with hewn stone, my paths hath he distorted' (Blayney), or made crooked (Eng. Ver.)

This building over him,—obstructing the cry of his voice,—blocking up the way with hewn stone like the masonry-blocking in the Pyramid passage,—and the passages so blocked being angular or crooked,—seems very significant of a pyramid.

It may further be noticed, that the nearness of Jeremiah, and more especially of Job, both in respect of date and place, to the building of the Pyramids, and the Pyramid's unparalleled greatness and notoriety—then, even much more than now—lends much force to any reasons above shown, in favour of this construction of their words.

NOTE ON ARTICLES 1 AND 2.

In replying to, or anticipating, critical objections (which are sure to be made), we must remember that the high probability *that the Pyramids are intended* in these texts, does not rest on the same word or root being used in each case; for the words are quite different. Neither does it rest on the roots of the words signifying anything characteristic of the Pyramid's form; such as pointedness, five-corneredness, convergency, etc. But that the other considerations above set forth strongly indicate Pyramids; and *that any alternative idea* is not so well supported; and that therefore any charge of fancifulness applies really to those who reject these in favour of any alternative ideas less supported.

ARTICLE 3.

(See page 541, vol. iii. of *Life and Work.*)

Zechariah iv. 7.—'Who art thou, O great mountain? before 'Zerubbabel thou shalt become מִישֹׁר (mee-shōhr, mas.) straight-lined, 'evenly fitted, adjusted, and smoothed: and he shall bring forth the 'head-stone thereof with shoutings, crying, Grace, grace, unto 'it!'

If we take the obvious connexion of the sentences,—rightly enough suggested by the word *thereof*—in the English version, and so understand 'the head-stone thereof' to mean the completing head-stone of the great mountain; the assertion that it shall become *a plain* (as in Eng. Ver.) is contradictory to the next assertion, that it shall be joyfully completed with a head-stone. The only alternative exegesis is, that the great mountain *means a difficulty* in the way of the completion of some desired edifice to which the head-stone refers: but this sense of mountain is not supported by the use of the word in other places of Scripture;[1] and if it were, it would still make the construction of the present sentence very unnatural, by omitting the essential object which the head-stone is to crown, and placing instead, the mountain to be abolished.

These alternatives being thus untenable, we turn to the original Hebrew, and we find that the root יָשַׁר (yāh-shar), he was straight or right,—from which the word (mee-shōhr) in the above text comes,—does not signify that which is *horizontal;* if it did, we should, confessedly, be limited to the inconsistent idea given in the Eng. Ver. But the word signifies that which is *even*, in the sense of not crooked, nor curved, serrated, nor having projections, as a pavement or stone facing with some of its tiles or stones sunk, and others above their proper level, or with their edges cropping out. The word imports just the idea that the Latin *rectus* does, in constituting the origin of our words *direct*, *correct*, *rectified*, and *right* morally (for, the Hebrew word is as freely used in a moral or figurative sense as is this, its equivalent in Latin and English), and right physically, as we say *a right line*, meaning, in the words of our geometrical definitions, 'a line that lies *evenly* between two points,' and, figuratively, right conduct.

[1] If we would avoid profitless and misleading speculation,—too common among commentators,—we must faithfully adhere to the following sound canon of interpretation. Admitting the Scriptures to have one Author, we must take its words in the sense taught in other places of the same Author's writings; and we must reject any symbolical meaning of merely man's devising, however plausible, for a word in one place of Scripture, in any case in which that word symbolizes differently in other places in Scripture; moreover, each material thing has generally but one symbolic sense, rarely more than two, throughout Scripture. For example, the word *mountain*, in Scripture, has a very distinct symbolic sense—Governmental authority, and—not the sense of great sins nor great difficulties, in Scripture, though these be continually so expressed by modern divines, to the detriment of true and sound exegesis. (See Part 4 of this article.)

PART 2.—*Examples.*

The preceding page has noted, that even when this Hebrew word in Scripture is used (as it frequently is) for a plain, it is not at all as a *necessarily horizontal* line or surface; but, as a *rectilineal*—and comparatively *smooth and even*—line or surface.

The following scriptural uses of this same form of the word (the masculine noun) illustrate this:—

Psalm xlv. 6.—'The sceptre of thy kingdom is *a straight* (right) sceptre.'

Isaiah xl. 4, and xlii. 16.—'The crooked shall be *made straight.*'

Psalm xxvi. 12.—'My foot standeth *in an even place.*'

So here, Zech. iv. 7.—It reads literally,—'Before the face of 'Zerubbabel (that is, under his superintendence), thou, great 'mountain, shalt become *adjusted, straight,* and *even.*'

The meaning is further illustrated in other forms of the same root in Scripture.[1] The following examples have important reference to the meaning intended in Zech. iv. 7. The texts are placed here, in such order as seems to illustrate and classify the senses in which the word is used.

Psalm xvii. 2.—'Thine eyes behold the things that are *equal,*' or *right* and *just.*

1 Samuel vi. 12.—'The kine took the *straight way.*'

Jeremiah xxxi. 9.—'I will cause thee to walk by the rivers of waters, in a *straight way* wherein they shall not stumble.'

An uneven or rugged way is the chief if not the only cause of stumbling here referred to; so that the sense of the verse shows that the word means *smooth and regular* in this text.

Judges xiv. 3.—'Samson said unto his father, Get her for me, for she *pleaseth* me *well,*'—literally,—she is *right* in mine eyes.

---

[1] In all such investigations as these, the *scriptural* usage,—the sense in which the word is used by the same Divine Author, is the point of special importance; far more than the usage of other authors, the secular and classical usage, even if it were different.

The word is in many other places translated *pleasing*, *righteous*, and UPRIGHT :—

Jeremiah xviii. 4.—'He made it again another vessel, as *seemed* '*good* to the potter to make it.'

Proverbs iii. 5, 6.—'Trust in the Lord with all thine heart; and 'lean not unto thine own understanding. In all thy ways acknow-'ledge Him, and He *shall direct* thy paths.'

Proverbs iv. 25.—'Let thine eyelids look *straight* before thee.'

2 Kings x. 3.—'Look even out the best and *meetest* [or fittest] of 'your master's sons,' and set him 'on his father's throne.'

Jeremiah xl. 5.—'The captain of the guard took Jeremiah, and 'said unto him . . . whither it seemeth good and *convenient* for 'thee to go, thither go . . . go back also to Gedaliah . . . and dwell 'with him among the people, or go wheresoever it seemeth *conveni-'ent* unto thee to go.'

But especially notice the following :—

Micah vii. 4.—'The best of them is a brier, *the most upright* is 'sharper than a thorn hedge.'

Ezekiel i. 23.—'Under the firmament, the wings of the living 'creatures were *straight*, the one toward the other; every one had 'two which covered on this side, and every one had two which 'covered on that side, their bodies.'

1 Kings vi. 35.—'He carved cherubim, and palm-trees, and open 'flowers, and covered them with gold FITTED upon the carved 'work.'

This last is the participial form of the word, *pual*, and therefore signifies intensity of the action; either,—as Gesenius suggests, the intensity of *flattening* the gold, by blows of a hammer to reduce it to a sheet,—or, it should rather be understood, the intensity or nicety of the *fitting* process mentioned, that is, to render its surface true and even with the curves and turns of the woodwork, so as to preserve *correctly* the form of the carved work thus encased. This is much more agreeable to the context than is Gesenius's notion; for, the *fitting* refers to the placing it on the carved work, rather than to the previous preparation of the gold in leaves. Besides, the carved work (cherubim, palm-trees, and open flowers) were *curved* and complex surfaces, not straight or flat ones.

This last instance of the use of the word (mee-shōhr) brings out very clearly, and in close connexion with our subject, the sense of accuracy of *fitting* and *adjustment* (irrespective of the primary sense, straightness) as a derived sense from the primary idea of straightness; this has led to its use for *evenness*, and thence *correctness* in general. But in the work of casing and finishing the Great Pyramid, the term is peculiarly appropriate both in its direct and derived sense.

From all the foregoing, the root is seen to express *straight*, *pleasing*, *corrected*, or *made upright*, EVEN-*fitted* (meaning closely and evenly), and *flush* with a given surface, as the technical English phrase goes. In view of this, the text in question would be 'under the direction and superintendence of Zerubbabel, thou, great mountain, shall become *straight-lined*, *evenly fitted*, *adjusted*, and *smoothed*,' and this word refers well to the completion of the Great Pyramid, by covering over its step-formed sides of roughly hewn blocks of the grey nummulitic limestone, with a smooth inclined surface of the compact and white Mokattam casing-stones with extraordinarily exact and close joints; and the word refers well to the polishing down of all inequalities, so as to produce its truly *even* or plane surfaces and straight arris-lines, giving to the mass its final and correct dimensions and angle.

PART 3.—*Impartial corroborative evidence* 200 *years* B.C. *Being the consistent opinion of the* LXX. *on the meaning of this Hebrew word in all places in the Scriptures.*

In corroboration of the above, we notice that the original word (mee-shōhr), was in this verse understood in a similar sense,—and not as a horizontal plain,—by the seventy chosen translators of the Greek version at Alexandria, 200 years before Christ.

Κατορθῶσαι is their expression. '*Thou shalt be thoroughly corrected or adjusted*' is the exact and literal sense of this Greek word, by which the Septuagint here render the Hebrew (mee-shōhr). Κατὰ in composition denotes here the intensity or thoroughness of the action.

But more than this. From their mode of translating the same Hebrew word in other places, we learn that their view of its primary sense was *not horizontalness*, but *straightness and evenness*, no matter whether vertical or inclined or horizontal. It is reassuring thus to perceive that those pre-Christian translators exactly concurred with the results of our independent critical examination of the usage of the word in our own codex of the Hebrew Scriptures.

This is seen by noticing that the LXX. had an appropriate Greek word, πεδεινή, which they habitually used in Scripture, in its accurate sense, to express *a plain* (straightness and evenness when combined *with horizontalness*), and accordingly, where they met with the Hebrew root, mee-shōhr, and its context decidedly *adds* to it the sense of horizontalness, they do, generally, render it by πεδεινή—*e.g.* Jer. xxi. 13, but not always, even in such cases; so little do they recognise the idea horizontalness as *necessarily* belonging to mee-shōhr—*e.g.*, 1 Kings xx. 23 and 25, πολεμήσωμεν κατ' εὐθὺ; and Joshua xiii. 9, 16, 17, 21, τὰς πόλεις τοῦ Μισώρ.

To the example in Jeremiah xxi. 13, we might have added Deut. iv. 43, ἐν τῇ γῇ τῇ πεδινῇ, and 2 Chron. xxvi. 10, etc., but Jeremiah is more to the point, for it was translated by the same commission as Zechariah, and therefore would illustrate their usage better than examples from the Pentateuch, which was translated into Greek by a different commission, about 100 years previously.

PART 4.—*The Context.*

The text, the translation of which is above criticised, Zech. iv. 7, has not only much *manifest* allusion to the Great Pyramid, but much other connexion therewith, not so obvious at first sight. Without here digressing to discuss the evidences of this, we pass on to notice a point too closely connected with the above criticism to be wholly remitted to a future paper. With such reference to the Great Pyramid, then, the symbolic word *mountain* (see foot-note, page 490) has a peculiar import in the text above criticised, as connected with other texts, *e.g.*, the stone which 'became a great mountain;' Daniel ii. 35. That Pyramid—the only one both original and purely non-idolatrous and non-subservient to this or that mortal's earthly glory like the rest of the pyramids,—that Pyramid, as far removed from polytheism as from atheism, from the afflicting atheism whose misguided votaries are still more egotistical and bent on visibly and ostentatiously magnifying themselves during their little earthly careers, than any other class of mankind,—that great Pyramid, the greatest of the stone cairns built by Divine command at various times, as tokens of God's past or future acts among men,—is set as a standing symbolic protest (of which 'the decree' in Psalm ii. is a brief verbal translation) *reserving the* GOVERNMENTAL *rights* of 'The King's Son,'—'Δυνάμεις μέλλοντος αἰῶνος,'—during the times in which the nations are left to walk *de facto* in their own ways, owning another lord,—'τοῦ σκότους τοῦ αἰῶνος τούτου.'

This part of the matter cannot be entirely passed over, though it must be only thus hinted at here, as it is a subject of kindred extent and importance with that to which this present volume is more expressly devoted.

APPENDIX 8.

ERRATA

IN 'LIFE AND WORK AT THE GREAT PYRAMID'

(SO FAR AS DISCOVERED UP TO APRIL 1868).

VOL. I.

Page	*Line*
159	6 from top, *for* no *read* not.
173	13 from top, *for* hard *read* broad.

VOL. II.

Page	*Line*
66	last but 1, *for* continuation *read* culmination.
126	*N.B.*—The distances entered for the 'trenches' from the base-sides of the G. Pyramid, refer to that building's present sides, which are roughly above 100 inches within the ancient casing-stone sides.
158	12 *ab imo*, *for* 334″ *read* 334°.
192	last but 1, *for* 5′ 0′ *read* 5′ 0″.
195	5 from top, *for* 18-inch rod *read* 18-foot rod.
458	Erase the third paragraph, and read the article on Sir Isaac Newton's value of the Sacred Cubit, following these pages of Errata.
458	2d line of par. 4, *for* Sir Isaac Newton *read* Sir Isaac Newton's numbers re-examined.

VOL. III.

Page	*Line*
135	18 from top, *for* realized;—*read* realized as a linear standard;—
135	last but 2, dele *nine*.
135	last but 1, *for* 25·07 *read* more than 25.
174	15 *ab imo*, *for* 0·85 *read* 2·85.
344	6 from top, *for* expressed *read* printed.
	Correct also the mistake of Mr. William Osburn having been thought to be dead. He is happily still living, and the author has received some very instructive letters from him, during the last six months.
505	2, of par. 3, *for* xi. 16 *read* xvi. 11.

SIR ISAAC NEWTON'S NUMBERS FOR THE LENGTH OF THE SACRED CUBIT OF THE HEBREWS.

AT p. 458, vol. ii. of *Life and Work at the Great Pyramid*,—there is an unfortunate misprinting of the calculated numbers representing in British inches, the quantities from which the mean '25·07, ± ·10 British inches,' for a new statement of the length of the above cubit, was derived. This *final* mean, is correctly given, as intended; so likewise are the *original* terms, expressed chiefly in Roman Unciæ, in Sir Isaac Newton's Dissertation on Cubits, reprinted at pp. 354-366. No important mischief therefore is likely to have accrued, from this error in printing one of the *intermediate* steps. But as the error is an undoubted blemish, which I much regret, have cancelled in the list of *errata*, and sincerely thank those who have called my attention to it,—I hasten to give the following discussion *de novo*.

At p. 365, of Sir Isaac's treatise above mentioned, he assumes $25\frac{3}{5}$ Homan unciæ, to represent the length of the sacred cubit of the Rebrews,—a cubit which he had elsewhere shown, there were grounds for believing that that people possessed before they went down into Egypt, and had had specially brought to their attention again, for religious matters, after leaving Egypt under Moses.

But Sir Isaac Newton was not confident of having obtained this cubit's precise length, to the last figure put down in his arithmetical expression. And he particularly and almost prophetically says,—

'This is what I thought proper to lay down *at present* with regard 'to the magnitude of this Cubit. Hereafter, perhaps those who 'shall view the sacred mount, and the monuments of the *Chaldæans*, 'by taking accurately the various dimensions of the stones, bricks, 'foundations, and walls, and comparing them together, will discover 'something *more certain and exact*.'

Now what Sir Isaac laid down at that then present time, was abundantly sufficient for his then purpose; or to prove, that there existed a most sensible and positive difference in the length of that sacred (or 25⅗ unciæ) cubit of the Hebrews, and of the profane cubit of the Egyptians;—whose length, expressed in the same Roman unciæ, was hardly more than 21·3. And in this last conclusion, he is so eminently borne out by all subsequent investigators, that that subject,—or the length of the *profane*, or Egyptian national, cubit,—need not be stirred again.

But within the last few years, another, and a more refined, or a residual question has arisen, which apparently never crossed Sir Isaac Newton's mind; viz. was the sacred cubit of the Hebrews, taken by itself, accurately the ten-millionth part of the length of the Polar semi-axis of the Earth? And as this quantity in Nature, according to modern science, is something very close to 25·8 Roman unciæ,—Sir Isaac's determination of 25 and ⅗, *i.e.* 25·6 of the same unciæ for the sacred cubit, is, to say the least of it, so near—especially for a confessed imperfect approximation, from a portion only of the materials collected,—that it becomes intensely important to submit *all* the data to a more rigid scrutiny than before; with the caution moreover in view, of assigning some limits, within which we may feel tolerably certain of the result.

The several quantities therefore, extracted from my reprint of Sir Isaac Newton's paper (but to which, in the original, I cordially refer all readers),—and reduced to British inches,—at the approximately assumed rate of 12·15 for 1 Attic foot; and 0·97 for 1 Roman *uncia*,—are as follows:—

No. of Approximation.	METHOD OF REFERENCE.	Attic Feet.	Roman Feet.	Roman Unciæ.	British Inches.	British Inches.
1.	From Talmudists and Josephus, by Greek cubit (p. 355, vol. ii.); between .	$2\frac{4}{7}$ and 2	... ...	... ...,	31·24 24·30	} 27·77
2.	From Talmudists, by proportions of human body (p. 355), between		... ...	$28\frac{4}{5}$ 24·	27·94 23·28	} 25·61
3.	Josephus by pillars of Temple (p. 356), between		$2\frac{1}{3}$ 2	28·0 24·0	27·16 23·28	} 25·22
4.	By Talmudists' and "all Jews'" ideas of a Sabbath-day's journey (p. 358), between . . .		$2\frac{1}{3}$ 2	28·0 24·0	27·16 23·28	} 25·22
5.	By Talmudists and Josephus, on steps to Inner Court (p. 358), between .			27·0 24·0	26·19 23·28	} 24·74
6.	By many Chaldaic and Hebrew proportions to Cubit of Memphis (p. 362), .			25·6	24·83	24·83
7.	From a statement by Mersennus, as to the length of a supposed copy of the Sacred Cubit of the Hebrews, preserved secretly (p. 363, vol. ii. *Life and Work*),			25·68	24·91	24·91

The simple mean of the last column = 25·47 British inches. But that is not a proper method there; because, not only has Sir Isaac Newton evidently shown that he had most confidence in his two last determinations; but his first, by its very wide limits, shows that it is by far the least trustworthy of all.

Some decrease of weight, therefore, for No. 1, and increase for Nos. 6 and 7, require to be made. How much precisely, it is impossible to say: but perhaps ½ for the former, and 3 for each of the two latter, the intermediate quantities being reckoned at 1 each,—may be considered fair and probable. In which case the mean comes out, 25·05 British inches. While, simply,—and in fact as I did on the first occasion, using then a slightly different value of the Roman uncia,—throwing away the on most objectionable observation, and taking a mean of the rest, unweighted, gives 25·09 of the same British inches.

But neither 25·09, nor 25·05 are fully safe, either in the second, or perhaps the first, place of decimals;—for—besides the uncertainty connected with the proper weighting of each of the results, according to the different kind of documentary evidence obtained by Sir Isaac Newton on each occasion,—there is considerable uncertainty in the value of a Roman uncia, expressed in British inches. We have assumed as above, that the former = 0·97 of the latter: but modern scientific and architectural authorities are found anywhere, between Zach at 0·9681 and Penrose at 0·97286; and might require us to reduce our final quantities by − ·05, or increase them by + ·06 of an inch; or by any intermediate figure.

Wherefore, the statement already printed at p. 458 of vol. ii. of *Life and Work at the Great Pyramid*,—i.e. 25·07 ± ·10 British inches, for the best result deducible from all Sir Isaac Newton's approved approximations for the length of the Sacred Cubit of the Hebrews,—is, if not as good a statement as can be made,—at least a great deal better than the 24·82 inches, absolute, which has been hitherto current in most English works; and beyond comparison better than the 20·7 inches, nearly, of the ORDNANCE SURVEY *Map of Jerusalem*.

This Ordnance quantity of 20·7 inches is evidently not the *sacred* cubit at all, but the *profane* cubit; and in the explanation of the scale at the foot of the above map, the revered names of 'Sacred' and 'Cubit of the Tabernacle' are given to precisely what Moses was so anxious to keep them from being confounded with, viz., the cubits of idolatrous Egypt and other Gentile nations; the inscription at one end of one of the Ordnance-Map scale-lines being '*Egyptian, Hebrew, Babylonian*,' and at the other end '*Royal or* '*Sacred cubits, also named cubits of the Tabernacle*.' If this map is one of those prepared, as believed by some, at the expense, and to the orders of, the Palestine Exploration Association,—such a radical error with regard to the *sacred cubit of the Hebrews* may well excite surprise. But if, on the contrary, the map is purely the work of the several Ordnance officers whose names are conspicuously engraved upon it,—the nation must regret that they should have so entirely ignored the researches of Sir Isaac Newton, the greatest philosopher their country ever produced, and in one of the most important of all questions that have ever been brought forward in either the science or history of metrical standards.

APPENDIX 9.

NOTE

INSERTED BY SPECIAL REQUEST OF A CONTRIBUTOR TO THIS APPENDIX,

ON THE QUESTION OF FURTHER EXPLORATION OF THE GREAT PYRAMID.

A FEW of the deeper readers of the foregoing work and Appendix, and of *Life and Work*, etc., will be impressed with the extreme importance of a further exploration of this unique structure, the Great Pyramid ; an exploration to be conducted with an *accuracy* and completeness worthy of the great interests now for the first time found to be involved,—palæologic, chronologic, metrologic, geodetic, geologic, astronomic, but above all, symbolic, that is to say, relating to the higher ideas intentionally embodied therein by its Originator, and for the sake of which the structure itself and its scientific facts were really designed ; according to the conviction of some of us, after a carefully impartial study of the subject with the present incomplete data.

In presence of the existing and increasing sense of the importance of the preservation, and of a further exploration of this pre-eminent 'sign and wonder,' a word of caution and guidance is here earnestly added. Most happily, the diplomatic as well as the pecuniary difficulties are great; and for this, our gratitude is due to the rulers of Egypt; for, if it had been otherwise, the structure would have been already so much injured by zealous but improper exploration, that the obtaining of the much-needed exact data would have been rendered hopeless. We can only wish that the governing powers had been equally firm in adopting the most stringent arrangements to arrest the destruction now continually in progress by the wantonness and perversity of crowds of tourists. The pecuniary and operative part of the work has likewise been a wholesome restraint on improper exploration. As bearing on this, we may state, that not far short of ten thousand pounds would be needed for such refined scientific work and engineering labours as would be properly required, including the due replacing, and re-enveloping of the parts to be exposed during the examination and measurement.

We say that these diplomatic and other restraints do, most happily, exist; for it would be worse than useless—a positive calamity—if such exploration were attempted at all—by persons not fully competent—scientifically, practically, instrumentally, and pecuniarily—and duly impressed with the serious moral responsibility of the work, and with the grave importance of devoting all skill and scientific means of extreme precision, to that part of the business which remains to be done, if it is attempted at all: seeing that the attempt will (as in Howard Vyse's case) necessarily expose fiducial parts, of inestimable value, to risks of irrecoverable obliteration by swarming tourists, and even by accidents and carelessness during the investigation.

Begging the esteemed author to insert this word of caution and suggestion, addressed to the worthy zeal for adequate completion of proofs or disproofs and data,—a zeal which his publications on the Great Pyramid are arousing, whether he wish it or not, in individuals, few and far between, yet collectively numerous,—I subscribe myself, his obedient servant,

WILLIAM PETRIE.

4 THE AVENUE, BLACKHEATH,
LONDON, S.E., *June* 2, 1868.

INDEX.

A

P

Q

R

LIFE AND WORK

AT

THE GREAT PYRAMID

During the Months of January, February, March, and April, A.D. 1865;

WITH A DISCUSSION OF THE FACTS ASCERTAINED:

By C. PIAZZI SMYTH, F.R.S.S. L. & E., F.R.A.S., F.R.S.S.A.,

HON. M., I.E., SCOT., P.S. ED., AND R.A.A.S. MUNICH AND PALERMO, PROFESSOR OF PRACTICAL ASTRONOMY IN THE UNIVERSITY OF EDINBURGH, AND ASTRONOMER-ROYAL FOR SCOTLAND.

In Three Vols. Demy 8vo, pp. 1653, Price 56s.

ILLUSTRATED WITH 38 PLATES AND SEVERAL WOODCUTS.

OPINIONS OF THE PRESS.

"Professor Piazzi Smyth deserves no common credit for his remarkable investigations; and we think he has fully made out a case for further inquiry. This, however, ought no longer to be left to private means, but developed by the resources of a powerful State; especially if that State be one which is about to change, without adequate consideration, its ancestral standards (which may prove to be older than any existing records) for an extremely artificial system of, at best, doubtful utility. Modern mechanical devices are capable of easily lifting and replacing without injury, the hugest stones in the Pyramid; and until that is done in the interior, and until the original sides are measured with Ordnance-Survey precision, there will always be a strong feeling that important attainable knowledge is being allowed to disappear gradually from the earth for want of a little liberality and decision."—*North British Review*, No. XCIII., *September* 1867.

"The Egyptians, upwards of 4000 or 5000 years ago, must have possessed a complete metrologic system, more perfect, we say it, than the metrical system of which we (the French) are so proud, and they have engraved the indelible remembrance of it on this mountain of stone."—*Revue des Deux Mondes, Paris*, 15*th October* 1867.

"Photography and mathematical instruments have made Pyramidology quite a fireside study."—*Newcastle Daily Journal*, 4*th May* 1868.

"Professor Smyth is highly qualified for the task of measurement; and he has measured with great diligence. . . . We see also many samples of power, and of great earnestness combined with careful industry."—*The Athenæum*, 20*th July* 1867.

"The conscientious care and accuracy of Professor Smyth's observations, the abundance of facts and figures embodied in these volumes, fairly challenge consideration and inquiry."—*The Scotsman*, 7*th November* 1867.

"Shortly after the publication of this pioneer volume (*Our Inheritance in the Great Pyramid*, 1864), the eminently qualified writer, following the chivalrous example of Greaves, went to Egypt to make his own observations on the Pyramid. The results are now given to the world in volumes which will certainly form an era in the history of Pyramid discovery."—*The Guardian*, 6*th November* 1867.

"As a record of pious devotion to science, and of labours undertaken from the loftiest motive by which man can be inspired,—a love of truth for the sake of it, and a determination to get at it,—these magnificent volumes will ever have a moral charm. . . . His conclusions it is difficult to accept. It is difficult also to reject them."—*Daily Review*, 2*d December* 1867.

"Few narratives of Egyptian travel are more entertaining; and we cannot fail to sympathize heartily with the author's zealous desire to pluck out the heart of this huge mystery of forty centuries ago, or to admire the spirit of enterprise, the courage, industry, and practical ingenuity which he displayed in so toilsome a task, entirely carried on at his own private expense for the advancement of scientific and antiquarian truth."—*The Illustrated London News*, 8*th June* 1867.

"These volumes are in their way unique, and well worthy the attention of those who devote themselves to the useful as well as the pleasant in science."—*The Standard*, 12*th June* 1857.

"The work, as a whole, is a marvel of careful accumulation of fact, and it is replete with the most succinct argument."—*The Globe, 10th June* 1867.

"In spite of the formidable appearance of three arge volumes, no one will ever look upon the time devoted to it as anything but well spent."—*The Church Standard, 18th May* 1867.

"The deep scientific and historical researches of these volumes is indeed refreshing, amid the increasing audacity of modern scepticism."—*The Voice upon the Mountains, May* 1867.

"The author of these volumes is one of those thoroughly energetic men, whom difficulties do not deter from. research. . . . The results of that examination (of the Great Pyramid), we have no hesitation in saying, are invaluable."—*Bell's Weekly Messenger, 27th April* 1867.

"*Life and Work at the Great Pyramid* is a monument of labour, of erudition, and of research, which may well challenge the admiration and gratitude of all who are interested in solving the mystery of that stupendous structure, which is rightly designated the greatest wonder of the world."— *The Morning Star, 15th April* 1867.

"We heartily recommend this book to the notice of all, especially scientific men."—*The Manchester Courier, April 16th*, 1867.
